# Differential Equations in Engineering

## Mathematical Engineering, Manufacturing, and Management Sciences

*Series Editor*
*Mangey Ram*
*Professor, Assistant Dean (International Affairs), Department of Mathematics, Graphic Era University, Dehradun, India*

The aim of this new book series is to publish the research studies and articles that bring up the latest development and research applied to mathematics and its applications in the manufacturing and management sciences areas. Mathematical tool and techniques are the strength of engineering sciences. They form the common foundation of all novel disciplines as engineering evolves and develops. The series will include a comprehensive range of applied mathematics and its application in engineering areas such as optimization techniques, mathematical modelling and simulation, stochastic processes and systems engineering, safety-critical system performance, system safety, system security, high assurance software architecture and design, mathematical modelling in environmental safety sciences, finite element methods, differential equations, reliability engineering, etc.

**Partial Differential Equations: An Introduction**
*Nita H. Shah and Mrudul Y. Jani*

**Linear Transformation**
Examples and Solutions
*Nita H. Shah and Urmila B. Chaudhari*

**Matrix and Determinant**
Fundamentals and Applications
*Nita H. Shah and Foram A. Thakkar*

**Non-Linear Programming**
A Basic Introduction
*Nita H. Shah and Poonam Prakash Mishra*

**Applied Soft Computing and Embedded System Applications in Solar Energy**
*Rupendra Kumar Pachauri, J. K. Pandey, Abhishek Sharma, Om Prakash Nautiyal, Mange Ram*

**Differential Equations in Engineering**
Research and Applications
*Edited by Nupur Goyal, Piotr Kulczycki, and Mangey Ram*

**Sustainability in Industry 4.0**
Challenges and Remedies
*Edited by Shwetank Avikal, Amit Raj Singh, Mangey Ram*

**Applied Mathematical Modeling and Analysis in Renewable Energy**
*Edited by Manoj Sahni and Ritu Sahni*

For more information about this series, please visit: https://www.routledge.com/Mathematical-Engineering-Manufacturing-and-Management-Sciences/book-series/CRCMEMMS

# Differential Equations in Engineering

## Research and Applications

Edited by
Nupur Goyal, Piotr Kulczycki, and Mangey Ram

CRC Press
Taylor & Francis Group
Boca Raton London New York

CRC Press is an imprint of the
Taylor & Francis Group, an **informa** business

First edition published 2022
by CRC Press
6000 Broken Sound Parkway NW, Suite 300, Boca Raton, FL 33487-2742

and by CRC Press
2 Park Square, Milton Park, Abingdon, Oxon, OX14 4RN



**Library of Congress Cataloging-in-Publication Data**

Names: Goyal, Nupur, editor. | Kulczycki, Piotr, editor. | Ram, Mangey, editor.
Title: Differential equations in engineering : research and applications / edited by Nupur Goyal, Piotr Kulczycki, and Mangey Ram.
Description: First edition. | Boca Raton, FL : CRC Press, 2022. | Series: Mathematical engineering, manufacturing, and management sciences | Includes bibliographical references and index.
Identifiers: LCCN 2021011041 (print) | LCCN 2021011042 (ebook) | ISBN 9780367613129 (hbk) | ISBN 9780367613143 (pbk) | ISBN 9781003105145 (ebk)
Subjects: LCSH: Engineering mathematics. | Differential equations.
Classification: LCC TA347.D45 D45 2022 (print) | LCC TA347.D45 (ebook) | DDC 620.001/51535--dc23
LC record available at https://lccn.loc.gov/2021011041
LC ebook record available at https://lccn.loc.gov/2021011042

ISBN: 978-0-367-61312-9 (hbk)
ISBN: 978-0-367-61314-3 (pbk)
ISBN: 978-1-003-10514-5 (ebk)

Typeset in Times
by Deanta Global Publishing Services, Chennai, India

# Contents

# Preface

Creating mathematical models of real objects constitutes a fundamental aspect of modern science and various practical applications. The classical tool, especially in the mechanical and electrical engineering disciplines, is provided here by the theory of differential equations. The resulting formal apparatus is the basis of many practical disciplines, e.g., optimal and robust control, and theory of chaos or dynamical systems. Other fields in which differential equations are widely used include broad areas of mathematics such as linear algebra, numerical methods, and many others. It is worth emphasizing the close, mutually inspiring connections between differential equations and physics. After mathematical statistics, differential equations are currently the most important discipline in applied mathematics, especially in engineering and economics, as well as recently in the natural sciences. They are of value to students and graduates in the above disciplines in recognizing and understanding the laws that govern the surrounding world.

This book touches upon a number of specific aspects of the theory and practice of differential equations. Such a breadth of subject matter relates to a large degree to the fact that the fundamental problems of differential equations have already been sufficiently widely described in the subject literature. The scope of this book concerns both linear issues, which often have analytical solutions, and nonlinear ones, frequently supported by numerical methods. The following topics are covered in this monograph:

- Element-free Galerkin methods in computational fracture mechanics.
- Evaporative capillary instability of swirling fluid layer with mass transfer.
- Control instruments of regularized problems based on mathematical modeling of structural perturbations with applications at the nodes of 25 bar truss systems.
- Numerical simulation of singularly perturbed differential equations with large delays, using exponential $B$-spline collocation method.
- Application of differential equations to understanding the instability of binary nanofluids.
- Mathematical modeling and analysis of prey-predator models.
- The incremental harmonic balance method for multiple degrees-of-freedom systems with time delays.
- Solution to the Dirac equation.
- Periodic solution of a nonlinear economic cycle model with a generic investment function.
- Response evolution of a marine riser in random sea waves.
- Solution of a system of partial differential equations governed by natural convective flow in a porous cavity.

At the end of each chapter there is a list of subject literature, facilitating further exploration of the individual topics.

We wish the readers a successful study of the material presented, leading to new inspiration, a deepening understanding of the described concepts, and also fruitful applications to the contemporary challenges of science and practice.

**Dr. Nupur Goyal**
**Prof. (D.Sc.) Piotr Kulczycki**
**Prof. (Dr.) Mangey Ram**

# Acknowledgments

We express our sincere thanks to the team at CRC Press, Taylor & Francis Group for giving us this opportunity and for support during this project. Particular gratitude is due to Ms. Cindy Renee Carelli, Executive Editor, CRC Press, Taylor & Francis Group, for the excellent help provided by her to us in the approval for this book. We would also like to acknowledge the help from Ms. Erin Harris, Senior Editorial Assistant, CRC Press, Taylor & Francis Group.

**Nupur Goyal**
**Piotr Kulczycki**
**Mangey Ram**

# Editors

**Nupur Goyal** earned her Bachelor's degree in Computer Science in 2009 from Kurukshetra University, Kurukshetra, Haryana, India. She earned her Master's degree in Mathematics in 2011 from H.N.B. Garhwal University, Srinagar, Uttarakhand, India and her Ph.D. degree from Graphic Era University, Dehradun, Uttarakhand, India in November 2016. Dr. Goyal's research interests are in the area of reliability theory and operations research. She was an Assistant Professor in the Mathematics Department of the Suraj Degree College, Mahendergarh, Haryana in 2016, an Assistant Professor in the Mathematics Department of the Garg Degree College, Laksar, Haridwar, India from 2017 to October 2018, and an Assistant Professor and Head of the Department of Applied Science and Humanities, Roorkee Institute of Technology, Roorkee, India from 2018 to January 2020. Currently, Dr. Goyal is an Assistant Professor in the Department of Mathematics, Graphic Era (Deemed to be University), Dehradun, India. She is a reviewer of various international journals by publishers such as Springer, Emerald, IEEE, and IJMEMS. Dr. Goyal has published more than 30 research papers in various reputed national and international journals, chapters in books published by Springer, Emerald, Taylor & Francis, Inderscience, etc., and has presented her research at national and international conferences. She has been the Guest Editor in special issues of a number of journals. Dr. Goyal is an Associate Editor of the *International Journal of Mathematical, Engineering and Management Sciences.* She is currently editing a book for CRC Press, Taylor & Francis Group. She has been a member of the organizing committees of a number of international and national conferences, seminars, and workshops.

**Piotr Kulczycki** graduated in Electrical Engineering from the AGH University of Science and Technology, Krakow, Poland, and in Applied Mathematics from the Jagiellonian University in Krakow. Professor Kulczycki currently holds the position of Professor at the Systems Research Institute of the Polish Academy of Sciences, where he is the Head of the Centre of Information Technology for Data Analysis Methods, as well as at the AGH University of Science and Technology, where he is the Head of the Division for Information Technology and Systems Research.

**Mangey Ram** earned his Ph.D. degree, with a major in Mathematics and a minor in Computer Science, from G.B. Pant University of Agriculture and Technology, Pantnagar, India. He has been a Faculty Member for 12 years and has taught several core courses in pure and applied mathematics at undergraduate, postgraduate, and doctorate levels. Dr. Ram is currently the Research Professor at Graphic Era (Deemed to be University), Dehradun, India. Before joining Graphic Era, he was a Deputy Manager (Probationary Officer) with the Syndicate Bank for a short period. Dr. Ram is Editor-in-Chief of *International Journal of Mathematical, Engineering and Management Sciences*, and *Journal of Reliability and Statistical Studies*, Editor-in-Chief of six book series published by Elsevier, CRC Press, Taylor & Francis

Group, Walter De Gruyter Publisher, and River Publisher, and has been the Guest Editor and member of the editorial board of various journals. Dr. Ram has published more than 225 research publications (journal articles, books, book chapters, conference articles) published by IEEE, Taylor & Francis, Springer, Elsevier, Emerald, World Scientific, and many other national and international journals and conferences. He has authored/edited more than 50 books with international publishers like Elsevier, Springer Nature, CRC Press, Taylor & Francis Group, Walter De Gruyter Publisher, and River Publisher. His fields of research are reliability theory and applied mathematics. Dr. Ram is a Senior Member of the IEEE, Senior Life Member of Operational Research Society of India, the Society for Reliability Engineering, Quality and Operations Management in India, and the Indian Society of Industrial and Applied Mathematics. Dr. Ram has been a member of the organizing committee of a number of international and national conferences, seminars, and workshops. He was conferred with the 'Young Scientist Award' awarded by the Uttarakhand State Council for Science and Technology, Dehradun, in 2009. He was awarded the 'Best Faculty Award' in 2011, the 'Research Excellence Award' in 2015, and most recently the 'Outstanding Researcher Award' in 2018 for his significant contributions to academia and research at the Graphic Era Deemed to be University, Dehradun, India.

# Contributors

**Arora Geeta**
School of Chemical Engineering and Physical Sciences, Lovely Professional University
Phagwara, India

**Geng Guoqian**
State Key Laboratory of Hydraulic Engineering Simulation and Safety, Tianjin University
Tianjin, China

**Zhu Haitao**
State Key Laboratory of Hydraulic Engineering Simulation and Safety, Tianjin University
Tianjin, China

**Zhao jun**
Faculty of International Business, School of English Studies, Tianjin Foreign Studies University
Tianjin, China

**Sharma Jyoti**
University Institute of Engineering and Technology, Panjab University
Chandigarh, India

**Vaid Mandeep Kaur**
Department of Mathematics, Lovely Professional University
Phagwara, India

**Mbock Koumbe**
Department of Mathematics and Physics, National Advanced School of Engineering
Yaounde, Cameroon

**Awasthi Mukesh Kumar**
Department of Mathematics, Babasaheb Bhimrao Ambedkar University
Lucknow, India

**Mitra Ranjan Kumar**
Department of Mechanical Engineering, National Institute of Technology
Durgapur, India

**Xu Lixin**
State Key Laboratory of Hydraulic Engineering Simulation and Safety, Tianjin University
Tianjin, China

**Pant Mohit**
Department of Mechanical Engineering, National Institute of Technology
Hamirpur, India

**Singh Payal**
Faculty of Technology & Engineering, Parul University
Vadodara, India

**Alam Pravez**
Department of Mathematics, Government Degree College
Gairsain, India

**Pandit Purnima**
Faculty of Technology & Engineering, M. S. University of Baroda
Vadodara, India

**Domingo Ayissi Raoul**
Department of Mathematics at Faculty of Sciences, University of Yaounde I
Yaounde, Cameroon

**Magloire Etoua Remy**
Department of Mathematics and Physics, National Advanced School of Engineering
Yaounde, Cameroon

**Asthana Rishi**
School of Engineering and Technology, BML Munjal University
Gurgaon, India

**Pandey S. K.**
Department of Mathematics, Sardar Patel University of Police Security and Criminal Justice
Jodhpur, India

**Kapoor Saurabh**
Department of Education in Science and Mathematics, Regional Institute of Education (NCERT)
Bhubaneswar, India

**Patel Tanvi**
Faculty of Technology & Engineering, Parul University
Vadodara, India

**Yu Yang**
State Key Laboratory of Hydraulic Engineering Simulation and Safety, Tianjin University
Tianjin, China

**Uddin Ziya**
School of Engineering and Technology, BML Munjal University
Gurgaon, India

# 1 Element-Free Galerkin Method for Computational Fracture Mechanics

*Mohit Pant*

## CONTENTS

## 1.1 INTRODUCTION

Differential equations form the basis of modeling of almost all phenomena in natural, physical, and biological sciences. Thereafter, numerical methods are employed to solve them and to find accurate solutions. In all approaches, the key feature of the numerical scheme is the conversion of a complex phenomenon to an easy representation with a mathematical equation, thereby allowing its solution using a computational machine, so that the problem can be simulated in a virtual environment, in order to satisfy the needs of analysis. Various numerical approaches have been established over the course of time, like the finite element method (FEM), the finite difference method (FDM), and the boundary element methods. Currently, FEM is employed to simulate a wide variety of problems in engineering. It is a robust and exhaustively developed method but has its own shortcomings.

For instance, in problems involving the simulation of large deformations, a significant fall in accuracy occurs because of the skewing or compressing of the underlying elements. Modeling of phase transformations and crack growths with arbitrary and complex paths are also cumbersome. Discretization, using typical meshing methodology in formulating these problems, raises problems in the handling of non-coinciding discontinuities (with the mesh boundaries). In a similar way, another area which poses a tough challenge to researchers is computational fracture mechanics, because of the inability to capture stress field oscillations with high accuracy near the crack tip region.

To tackle these drawbacks in conventional numerical techniques, meshfree methods have been widely proposed and exhaustively developed. Over the past thirty years, meshfree methods have become one of the most coveted methods for solving problems, ranging from astronomical problems to simulation of solid mechanics problems, fluid flow, and problems of modal analysis, heat transfer simulation, and optimization of numerical solutions.

## 1.2 HISTORICAL DEVELOPMENTS IN MESHFREE METHODS

The very first meshfree method was smoothed-particle hydrodynamics [1], which was capable of solving problems of fluid dynamics, heat conduction, simulation of machining, and solid mechanics with ease.

Whereas modified adaptations of smoothed-particle hydrodynamics were derived from the strong form, a few different methods based on the weak form were developed in the 1990s. The major use of these methods was in the area of solid mechanics. Using the partition of unity (PU) concept, enrichment of the displacement-based approximation was developed by Moës [2]. Belytschko and Black [3] devised the extended finite element method (XFEM) as an alternative to FEM. This exploits the advantages of the traditional finite element procedure by eliminating the requirement for remeshing, thus making it simpler than other meshfree methods for different applications. To eliminate the instabilities in smoothed-particle hydrodynamic methods, Liu [4] developed the meshfree local Petrov-Galerkin (MLPG) method and the reproducing kernel particle method [5]. MLPG is better than other meshfree methods as there is no need for dummy elements, as in the element-free Galerkin method (EFGM), and hence no specific technique for integration is required. The dynamic crack propagation was simulated using the cracking particles method (CPM) [5] but, without the desired accuracy, the method was not as widely accepted as were other meshfree methods.

Compared with all the major meshfree methods, EFGM has contributed the most toward simulating problems of applied mechanics. The EFGM [6, 7] is a meshfree method formulated in 1994 and uses the weak form universally in all its computations. A number of applications of EFGM can be seen in the field of fracture mechanics [8–11] and it has emerged as a successful tool to solve a range of problems.

Shape function formulation, with Lagrange's multiplier technique [8, 12] and the moving least square (MLS) approximation [9], were some of the earliest modifications to the EFGM. Enforcement of boundary conditions was accomplished using the penalty method [13] that ensured the positive-banded definite equations. For shape function construction, a blend of the MLS approach with the radial basis function [14] was utilized, but the method failed to achieve accurate results.

## 1.3 ELEMENT-FREE GALERKIN METHOD

Being a meshfree method, the EFGM requires definition of only the nodal data in the domain, in conjunction with necessary boundary conditions, to obtain the approximation function. This eliminates the need for element connectivity data, as in the finite element method (FEM). As a characteristic feature, the same space is deployed to construct both test and trial functions, using moving least square (MLS) approximants.

MLS approximants are constructed by three components, defined as the compact spot linked to each node, the location of dependent coefficients, and the polynomial basis. The support of the weight function determines its domain of influence. The connectivity of the nodes is defined by overlap of the domain of influence of the node.

## 1.4 MOVING LEAST SQUARE (MLS) APPROXIMATIONS

In EFGM, MLS approximates to the unknown function $T(\mathbf{x})$, $T^h(\mathbf{x})$ [7] as

$$u^h(\mathbf{x}) = \sum_{j=1}^{m} c_j(\mathbf{x}) b_j(\mathbf{x}) = c^T(\mathbf{x})\,\mathbf{b}(\mathbf{x}) \tag{1.1}$$

where

$\mathbf{c}(\mathbf{x})$ is the basis functions defined as:

$$\mathbf{c}^T(\mathbf{x}) = [1,\ x,\ y,\ z,\ xy,\ yz,\ zx, \ldots\ x^{k'},\ y^{k'},\ z^{k'}] \tag{1.2}$$

and $\mathbf{b}(\mathbf{x})$ indicates the unknown coefficients for vector

$$\mathbf{b}^T(\mathbf{x}) = \left[b_1(\mathbf{x}), b_2(\mathbf{x}), b_3(\mathbf{x}), \ldots b_m(\mathbf{x})\right] \tag{1.3}$$

where $\mathbf{x}^T = [x\ y\ z]$,
$k'$ = degree of the polynomial
$m$ = number of terms in the basis.

1-D: Linear basis

$$\mathbf{c}^T(\mathbf{x}) = [1, x], \quad (m = 2, \text{linear}) \tag{1.4}$$

$$\mathbf{b}^T(\mathbf{x}) = \left[b_1(\mathbf{x}), b_2(\mathbf{x})\right] \tag{1.5}$$

Quadratic basis

$$\mathbf{c}^T(\mathbf{x}) = [1, x, x^2], \quad (m = 3, \text{quadratic}) \tag{1.6}$$

$$\mathbf{b}^T(\mathrm{x}) = \left[b_1(\mathbf{x}), b_2(\mathbf{x}), b_3(\mathbf{x})\right] \tag{1.7}$$

2-D: Linear basis

$$\mathbf{c}^T(\mathbf{x}) = [1, x, y], \quad (m = 3, \text{linear}) \tag{1.8}$$

$$\mathbf{b}^T(\mathbf{x}) = \left[b_1(\mathbf{x}), b_2(\mathbf{x}), b_3(\mathbf{x})\right] \tag{1.9}$$

Quadratic basis

$$\mathbf{c}^T(\mathbf{x}) = \left[1, x, y, xy, x^2, \mathrm{y}^2\right], \tag{1.10}$$

$$\mathbf{b}^T(\mathbf{x}) = \left[b_1(\mathbf{x}), b_2(\mathbf{x}), b_3(\mathbf{x}), b_4(\mathbf{x}), b_5(\mathbf{x}), b_6(\mathbf{x})\right] \tag{1.11}$$

Similarly, 3-D: Linear basis will be given as

$$\mathbf{c}^T(\mathbf{x}) = [1, x, y, z] \tag{1.12}$$

$$\mathbf{b}^T(\mathrm{x}) = \left[b_1(\mathrm{x}), b_2(\mathrm{x}), b_3(\mathrm{x}), b_4(\mathrm{x})\right] \tag{1.13}$$

while quadratic basis is given as

$$\mathbf{c}^T(\mathbf{x}) = \left[1,\ x,\ y,\ xy,\ yz,\ zx,\ x^2,\ y^2,\ z^2\right], \tag{1.14}$$

$$\mathbf{b}^T(\mathbf{x}) = \left[b_1(\mathbf{x}), b_2(\mathbf{x}), b_3(\mathbf{x}), b_4(\mathbf{x}), b_5(\mathbf{x}), b_6(\mathbf{x}), b_7(\mathbf{x}), b_8(\mathbf{x}), b_9(\mathbf{x}), b_{10}(\mathbf{x})\right] \tag{1.15}$$

The coefficients $\mathbf{b}(\mathbf{x})$ in Eq. (1.1) are the functions of $\mathbf{x}$ and $\mathbf{b}(\mathbf{x})$ at any point $\mathbf{x}$ and are obtained by minimizing the weighted least square sum (i.e., the least square sum in a weighted form between local approximation and the nodal parameter $T_I$).

$$L(\mathbf{x}) = \sum_{I=1}^{n} w\left(\mathbf{x}-\mathbf{x}_I\right)\left[\mathbf{c}^T(\mathbf{x})\mathbf{b}(\mathbf{x}) - T_I\right]^2 \tag{1.16}$$

where $T_I$ denotes the parameter linked with node $I$. This is clearly represented in Figure 1.1, where $w(\mathbf{x}-\mathbf{x}_I)$ denotes weight function with compact support linked to node $I$, and $n$ denotes the number of nodes within the influence domain containing point $\mathbf{x}$, i.e., $w(\mathbf{x} - \mathbf{x}_I) \neq 0$ as shown in Figures 1.2–1.3.

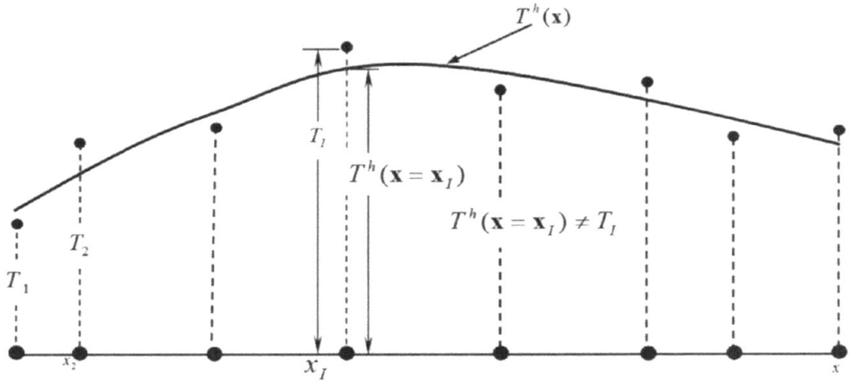


**FIGURE 1.1** Difference between $T_I$ and $T^h(\mathbf{x})$.

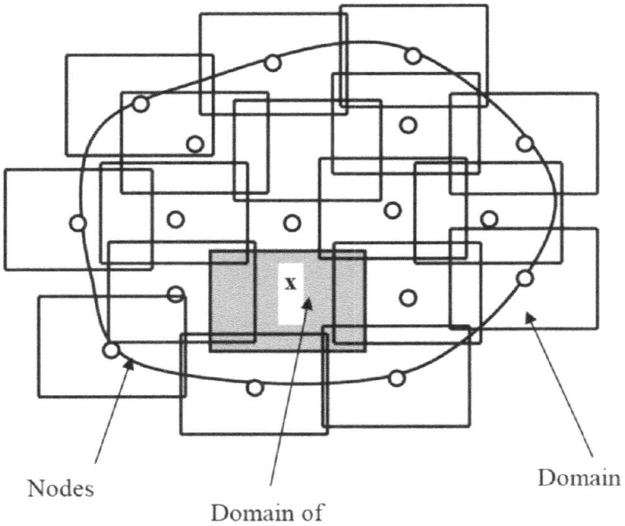


**FIGURE 1.2** Rectangular domain of influence in two-dimensional domains.

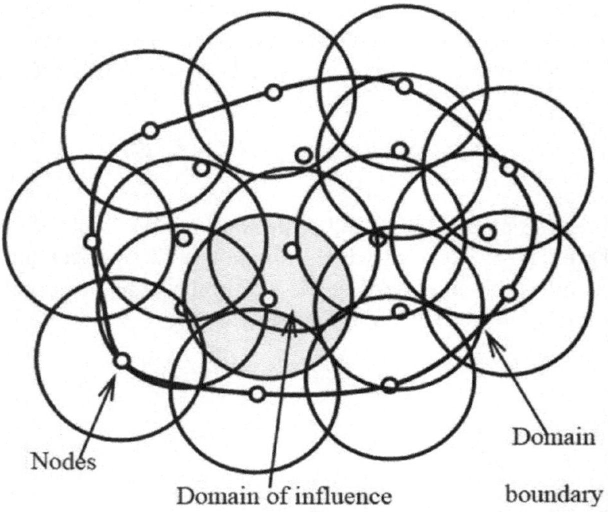


**FIGURE 1.3** Circular domain of influence in two dimensions, using tensor product domains.

The fixed value of $L$ given by Eq. (1.16) in comparison to $\mathbf{b(x)}$ gives us a set of linear equations:

$$\mathbf{A(x)\,b(x) = B(x)T} \tag{1.17}$$

or

$$\mathbf{b(x) = A^{-1}(x)\,B(x)T} \tag{1.18}$$

For 1-D:

$$\begin{aligned}\mathbf{A}(x) &= \sum_{i=1}^{n} w(x-x_I)c(x_I)c^T(x_I)\\ &= w(x-x_1)\begin{bmatrix}1 & x_1\\ x_1 & x_1^{\,2}\end{bmatrix}\\ &\quad+\ldots\ldots\ldots+w(x-x_n)\begin{bmatrix}1 & x_n\\ x_n & x_n^{\,2}\end{bmatrix}\end{aligned} \tag{1.19}$$

$$\begin{aligned}\mathrm{B}(x) &= \left[w(x-x_1)\mathrm{c}(x_1),\ldots,w(x-x_n)\mathrm{c}(x_n)\right]\\ &= \left\{w(x-x_1)\begin{bmatrix}1\\ x_1\end{bmatrix},\ldots\ldots,w(x-x_n)\begin{bmatrix}1\\ x_n\end{bmatrix}\right\}\end{aligned} \tag{1.20}$$

For 2-D:

$$\mathbf{A} = \sum_{I=1}^{n} w(\mathbf{x}-\mathbf{x}_I)\mathbf{c}(\mathbf{x}_I)\mathbf{c}^T(\mathbf{x}_I)$$

$$= w(\mathbf{x}-\mathbf{x}_1)\begin{bmatrix} 1 & x_1 & y_1 \\ x_1 & x_1^2 & x_1y_1 \\ y_1 & x_1y_1 & y_1^2 \end{bmatrix} + w(\mathbf{x}-\mathbf{x}_2)\begin{bmatrix} 1 & x_2 & y_2 \\ x_2 & x_2^2 & x_2y_2 \\ y_2 & x_2y_2 & y_2^2 \end{bmatrix} + \ldots\ldots + w(\mathbf{x}-\mathbf{x}_n)\begin{bmatrix} 1 & x_n & y_n \\ x_n & x_n^2 & x_ny_n \\ y_n & x_ny_n & y_n^2 \end{bmatrix} \tag{1.21}$$

$$\mathbf{B}(\mathbf{x}) = \{w(\mathbf{x}-\mathbf{x}_1)\mathbf{c}(\mathbf{x}_1), w(\mathbf{x}-\mathbf{x}_2)\mathbf{c}(\mathbf{x}_2), \ldots\ldots\ldots\ldots\ldots\ldots, w(\mathbf{x}-\mathbf{x}_n)\mathbf{c}(\mathbf{x}_n)\}$$
$$= \left\{ w(\mathbf{x}-\mathbf{x}_1)\begin{bmatrix} 1 \\ x_1 \\ y_1 \end{bmatrix}, w(\mathbf{x}-\mathbf{x}_2)\begin{bmatrix} 1 \\ x_2 \\ y_2 \end{bmatrix}, \ldots\ldots\ldots\ldots\ldots\ldots\ldots\ldots, w(\mathbf{x}-\mathbf{x}_n)\begin{bmatrix} 1 \\ x_n \\ y_n \end{bmatrix} \right\} \tag{1.22}$$

For 3-D:

$$\mathbf{A}(\mathbf{x}) = \sum_{I=1}^{n} w(\mathbf{x}-\mathbf{x}_I)\mathbf{c}(\mathbf{x}_I)\mathbf{c}^T(\mathbf{x}_I)$$

$$= w(\mathbf{x}-\mathbf{x}_1)\begin{bmatrix} 1 & x_1 & y_1 & z_1 \\ x_1 & x_1^2 & x_1y_1 & x_1z_1 \\ y_1 & x_1y_1 & y_1^2 & y_1z_1 \\ z_1 & x_1z_1 & y_1z_1 & z_1^2 \end{bmatrix} + w(\mathbf{x}-\mathbf{x}_2)\begin{bmatrix} 1 & x_2 & y_2 & z_2 \\ x_2 & x_2^2 & x_2y_2 & x_2z_2 \\ y_2 & x_2y_2 & y_2^2 & y_2z_2 \\ z_2 & x_2z_2 & y_2z_2 & z_2^2 \end{bmatrix} + \cdots + +w(\mathbf{x}-\mathbf{x}_n)\begin{bmatrix} 1 & x_n & y_n & z_n \\ x_n & x_n^2 & x_ny_n & x_nz_n \\ y_n & x_ny_n & y_n^2 & y_nz_n \\ z_n & x_nz_n & y_nz_n & z_n^2 \end{bmatrix} \tag{1.23}$$

$$\mathbf{B}(\mathbf{x}) = \left\{ w(\mathbf{x}-\mathbf{x}_1)\mathbf{c}(\mathbf{x}_1), w(\mathbf{x}-\mathbf{x}_2)\mathbf{c}(\mathbf{x}_2), \ldots\ldots\ldots\ldots\ldots\ldots, w(\mathbf{x}-\mathbf{x}_n)\mathbf{c}(\mathbf{x}_n) \right\}$$

$$= \left\{ w(\mathbf{x}-\mathbf{x}_1)\begin{bmatrix} 1 \\ x_1 \\ y_1 \\ z_1 \end{bmatrix}, w(\mathbf{x}-\mathbf{x}_2)\begin{bmatrix} 1 \\ x_2 \\ y_2 \\ z_2 \end{bmatrix}, \ldots\ldots\ldots\ldots\ldots\ldots, w(\mathbf{x}-\mathbf{x}_n)\begin{bmatrix} 1 \\ x_n \\ y_n \\ z_n \end{bmatrix} \right\} \tag{1.24}$$

By substituting Eq. (1.18) in Eq. (1.1), the MLS approximant is obtained as:

$$T^h(\mathbf{x}) = \sum_{I=1}^{n} \Phi_I(\mathbf{x}) T_I = \Phi^T(\mathbf{x})\mathbf{T} \tag{1.25}$$

where

$$\Phi^T(\mathrm{x}) = \left\{ \Phi_1(\mathrm{x}), \Phi_2(\mathrm{x}), \Phi_3(\mathrm{x}), \ldots \Phi_n(\mathrm{x}) \right\} \tag{1.26}$$

$$\mathbf{T}^T = \begin{bmatrix} T_1, & T_2, & T_3 & \ldots & T_n \end{bmatrix} \tag{1.27}$$

$\Phi_I(\mathbf{x})$ (meshfree shape function) is defined as:

$$\Phi_I(\mathbf{x}) = \sum_{j=1}^{m} c_j(\mathbf{x}) \left( \mathbf{A}^{-1}(\mathbf{x})\mathbf{B}(\mathbf{x}) \right)_{jI} = \mathbf{c}^T \mathbf{A}^{-1} \mathbf{B}_I \tag{1.28}$$

The partition of unity requirements for the shape function $\Phi_I(\mathbf{x})$ [15] are given as:

$$\sum_{I=1}^{n} \Phi_I(\mathbf{x}) = 1 \tag{1.29a}$$

$$\sum_{I=1}^{n} \Phi_I(\mathbf{x}) x_I = x \tag{1.29b}$$

$$\sum_{I=1}^{n} \Phi_I(\mathbf{x}) y_I = y \tag{1.29c}$$

$$\sum_{I=1}^{n} \Phi_I(\mathbf{x}) z_I = z \tag{1.29d}$$

The derivative of the shape function is calculated as:

$$\Phi_{I,\mathrm{x}}(\mathbf{x}) = (\mathbf{c}^T \mathbf{A}^{-1} \mathbf{B}_I)_{,\mathrm{x}} = \mathbf{c}^T{}_{,\mathrm{x}} \mathbf{A}^{-1} \mathbf{B}_I + \mathbf{c}^T (\mathbf{A}^{-1})_{,\mathrm{x}} \mathbf{B}_I + \mathbf{c}^T \mathbf{A}^{-1} \mathbf{B}_{I,\mathrm{x}} \tag{1.30}$$

where

$$\mathbf{B}_{I,\mathbf{x}}(\mathbf{x}) = \frac{dw}{d\mathbf{x}}(\mathbf{x}-\mathbf{x}_I)\mathbf{c}(\mathbf{x}_I) \tag{1.31}$$

and $\left(\mathbf{A}^{-1}{}_{,\mathbf{x}}\right)$ is computed as:

$$\mathbf{A}^{-1}{}_{,\mathbf{x}} = -\mathbf{A}^{-1}\mathbf{A}_{,\mathbf{x}}\,\mathbf{A}^{-1} \tag{1.32}$$

where $\mathbf{A}_{,\mathbf{x}} = \sum_{I=1}^{n} \frac{dw}{d\mathbf{x}}(\mathbf{x}-\mathbf{x}_I)\mathbf{c}(\mathbf{x}_I)\mathbf{c}^T(\mathbf{x}_I)$

## 1.5 EFFICIENT CALCULATION OF THE SHAPE FUNCTION

It is essential to calculate the inverse of the moment matrix, i.e., $\mathbf{A}^{-1}$, in order to calculate the meshfree shape functions $\Phi_I$. For 1-D problems this matrix inversion is not difficult, but in problems of higher dimension, this becomes computationally tedious. To overcome this, an alternative approach was developed [16], involving the lower–upper (**LU**) decomposition of the **A** matrix. The shape function is given as:

$$\Phi_I(\mathbf{x}) = \mathbf{c}^T(\mathbf{x})\mathbf{A}^{-1}(\mathbf{x})\mathbf{B}_I(\mathbf{x}) = \beta^T(\mathbf{x})\mathbf{B}_I(\mathbf{x}) \tag{1.33}$$

where

$$\beta^T(\mathbf{x}) = \mathbf{c}^T(\mathbf{x})\mathbf{A}^{-1}(\mathbf{x}). \tag{1.34}$$

Thus,

$$\mathbf{A}(\mathbf{x})\,\beta(\mathbf{x}) = \mathbf{c}(\mathbf{x}) \tag{1.35}$$

The vector $\beta(\mathbf{x})$ is evaluated by **LU** decomposition, followed by back substitution.

The expression below gives the partial derivatives of $\beta(\mathbf{x})$:

$$\mathbf{A}(\mathbf{x})\beta_{,x}(\mathbf{x}) = \mathbf{c}_{,x}(\mathbf{x}) - \mathbf{A}_x(\mathbf{x})\beta(\mathbf{x}) \tag{1.36}$$

$$\mathbf{A}(\mathbf{x})\beta_{,y}(\mathbf{x}) = \mathbf{c}_{,y}(\mathbf{x}) - \mathbf{A}_y(\mathbf{x})\beta(\mathbf{x}) \tag{1.37}$$

$$\mathbf{A}(\mathbf{x})\beta_{,z}(\mathbf{x}) = \mathbf{c}_{,z}(\mathbf{x}) - \mathbf{A}_z(\mathbf{x})\beta(\mathbf{x}) \tag{1.38}$$

$$\mathbf{A}(\mathbf{x})\beta_{,xx}(\mathbf{x}) = \mathbf{c}_{,xx}(\mathbf{x}) - \mathbf{A}_{,xx}(\mathbf{x})\beta(\mathbf{x}) - 2\mathbf{A}_{,x}(\mathbf{x})\beta_{,x}(\mathbf{x}) \tag{1.39}$$

$$\mathbf{A}(\mathbf{x})\beta_{,yy}(\mathbf{x}) = \mathbf{c}_{,yy}(\mathbf{x}) - \mathbf{A}_{,yy}(\mathbf{x})\beta(\mathbf{x}) - 2\mathbf{A}_{,y}(\mathbf{x})\beta_{,y}(\mathbf{x}) \tag{1.40}$$

$$\mathbf{A}(\mathbf{x})\beta_{,zz}(\mathbf{x}) = \mathbf{c}_{,zz}(\mathbf{x}) - \mathbf{A}_{,zz}(\mathbf{x})\beta(\mathbf{x}) - 2\mathbf{A}_{,z}(\mathbf{x})\beta_{,z}(\mathbf{x}) \tag{1.41}$$

$$\mathbf{A}(\mathbf{x})\beta_{,xy}(\mathbf{x}) = \mathbf{c}_{,xy}(\mathbf{x}) - \mathbf{A}_{,xy}(\mathbf{x})\beta(\mathbf{x}) - \mathbf{A}_{,x}(\mathbf{x})\beta_{,y}(\mathbf{x}) - \mathbf{A}_{,y}(\mathbf{x})\beta_{,x}(\mathbf{x}) \quad (1.42)$$

$$\mathbf{A}(\mathbf{x})\beta_{,yz}(\mathbf{x}) = \mathbf{c}_{,yz}(\mathbf{x}) - \mathbf{A}_{,yz}(\mathbf{x})\beta(\mathbf{x}) - \mathbf{A}_{,y}(\mathbf{x})\beta_{,z}(\mathbf{x}) - \mathbf{A}_{,z}(\mathbf{x})\beta_{,y}(\mathbf{x}) \quad (1.43)$$

$$\mathbf{A}(\mathbf{x})\beta_{,zx}(\mathbf{x}) = \mathbf{c}_{,zx}(\mathbf{x}) - \mathbf{A}_{,zx}(\mathbf{x})\beta(\mathbf{x}) - \mathbf{A}_{,z}(\mathbf{x})\beta_{,x}(\mathbf{x}) - \mathbf{A}_{,x}(\mathbf{x})\beta_{,z}(\mathbf{x}) \quad (1.44)$$

The derivatives of shape function are given as:

$$\Phi_{I,x}(\mathbf{x}) = \beta^T{}_{,x}(\mathbf{x})\mathbf{B}_I(\mathbf{x}) + \beta^T(\mathbf{x})\mathbf{B}_{I,x}(\mathbf{x}) \quad (1.45)$$

$$\Phi_{I,y}(\mathbf{x}) = \beta^T{}_{,y}(\mathbf{x})\mathbf{B}_I(\mathbf{x}) + \beta^T(\mathbf{x})\mathbf{B}_{I,y}(\mathbf{x}) \quad (1.46)$$

$$\Phi_{I,z}(\mathbf{x}) = \beta^T{}_{,z}(\mathbf{x})\mathbf{B}_I(\mathbf{x}) + \beta^T(\mathbf{x})\mathbf{B}_{I,z}(\mathbf{x}) \quad (1.47)$$

$$\Phi_{I,xx}(\mathbf{x}) = \beta^T{}_{,xx}(\mathbf{x})\mathbf{B}_I(\mathbf{x}) + 2\beta^T{}_{,x}(\mathbf{x})\mathbf{B}_{I,x}(\mathbf{x}) + \beta^T(\mathbf{x})\mathbf{B}_{I,xx}(\mathbf{x}) \quad (1.48)$$

$$\Phi_{I,yy}(\mathbf{x}) = \beta^T{}_{,yy}(\mathbf{x})\mathbf{B}_I(\mathbf{x}) + 2\beta^T{}_{,y}(\mathbf{x})\mathbf{B}_{I,y}(\mathbf{x}) + \beta^T(\mathbf{x})\mathbf{B}_{I,yy}(\mathbf{x}) \quad (1.49)$$

$$\Phi_{I,zz}(\mathbf{x}) = \beta^T{}_{,zz}(\mathbf{x})\mathbf{B}_I(\mathbf{x}) + 2\beta^T{}_{,z}(\mathbf{x})\mathbf{B}_{I,z}(\mathbf{x}) + \beta^T(\mathbf{x})\mathbf{B}_{I,zz}(\mathbf{x}) \quad (1.50)$$

$$\Phi_{I,xy}(\mathbf{x}) = \beta^T{}_{,x}(\mathbf{x})\mathbf{B}_{I,y}(\mathbf{x}) + \beta^T{}_{,xy}(\mathbf{x})\mathbf{B}_I(\mathbf{x}) + \beta^T(\mathbf{x})\mathbf{B}_{I,xy}(\mathbf{x}) + \beta^T{}_{,y}(\mathbf{x})\mathbf{B}_{I,x}(\mathbf{x}) \quad (1.51)$$

$$\Phi_{I,yz}(\mathbf{x}) = \beta^T{}_{,y}(\mathbf{x})\mathbf{B}_{I,z}(\mathbf{x}) + \beta^T{}_{,yz}(\mathbf{x})\mathbf{B}_I(\mathbf{x}) + \beta^T(\mathbf{x})\mathbf{B}_{I,yz}(\mathbf{x}) + \beta^T{}_{,z}(\mathbf{x})\mathbf{B}_{I,y}(\mathbf{x}) \quad (1.52)$$

$$\Phi_{I,zx}(\mathbf{x}) = \beta^T{}_{,z}(\mathbf{x})\mathbf{B}_{I,x}(\mathbf{x}) + \beta^T{}_{,zx}(\mathbf{x})\mathbf{B}_I(\mathbf{x}) + \beta^T(\mathbf{x})\mathbf{B}_{I,zx}(\mathbf{x}) + \beta^T{}_{,x}(\mathbf{x})\mathbf{B}_{I,z}(\mathbf{x}) \quad (1.53)$$

## 1.6 WEIGHT FUNCTION

The final approximation $T^h(\mathbf{x})$ of EFGM and other meshfree techniques depends on the choice of weight function selected. Therefore, an appropriate weight function selection plays a crucial role in all these methods. Here the weight function has a non-zero value in the influence domain of the node. The shape function $\mathbf{\Phi}_I$ inherits the continuity and smoothness of the weight function $w(\mathbf{x} - \mathbf{x}_I)$. It is illustrated in Figure 1.4.

The typical weight function should ensure the following conditions:

- In the domain of influence, it must have a positive value, should be differentiable, and continuous.
- The magnitude of the weight function should follow a decrease in magnitude with increasing distance between $\mathbf{x}$ to $\mathbf{x}_I$.
- Outside the domain of influence, the weight function should be zero.
- The weight function must produce a greater magnitude for a node nearer to it than for one farther away.
- The total number of nodes in the influence domain must exceed the total count terms in the basis function ($n > m$).

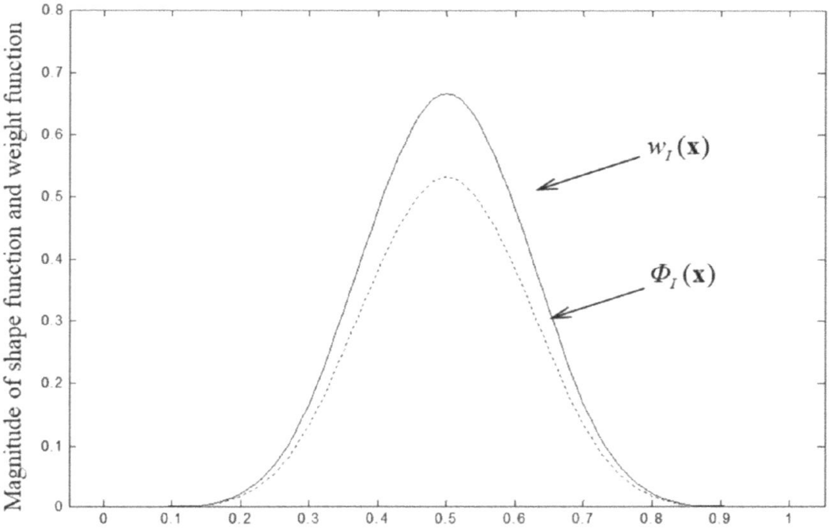


**FIGURE 1.4** A plot of weight function *versus* the corresponding shape function.

In this work, the exponential weight function has been considered for simulation due to its accuracy.

$$f(r) \equiv \begin{Bmatrix} 100^{-r} & 0 \le r \le 1 \\ 0 & r > 1 \end{Bmatrix} \tag{1.54}$$

For a parallelepiped domain of influence, the weight function for point $\mathbf{x}$ is computed in the following way:

$$w(\mathbf{x}-\mathbf{x}_I) = f\left(\left|x-x_I\right| / d_{mxI}\right) f\left(\left|y-y_I\right| / d_{myI}\right) f\left(\left|z-z_I\right| / d_{mzI}\right) \tag{1.55}$$

where $d_{mxI}=d_{\max}\ c_{xI}\ d_{myI}=d_{\max}\ c_{yI}\ d_{mzI}=d_{\max}\ c_{zI}$ and $d_{\max}$ are the scaling parameters, and $c_{xI}$, $c_{yI}$ and $c_{zI}$ represent the distances to the nearest neighbors denoted by $c_{xI} = \max_j \left|x_I - x_j\right|$, $c_{yI} = \max_j \left|y_I - y_j\right|$, $c_{zI} = \max_j \left|z_I - z_j\right|$.

The weight function's derivatives are obtained as:

$$w_x = \frac{dw_x}{dx} w_y w_z,\ w_y = \frac{dw_y}{dx} w_x w_z \text{ and } w_z = \frac{dw_z}{dx} w_x w_y \tag{1.56}$$

## 1.7 NUMERICAL INTEGRATION

Evaluation of the conductivity matrix (**K**) and the force vector (**f**) needs a process of integration within the domain, equivalent to performing an area integration within two dimension numerical integration technique such as Gauss quadrature being utilized for integration.

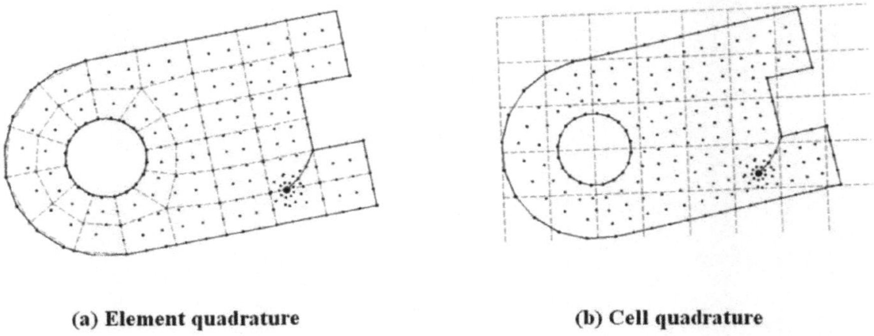


**FIGURE 1.5** Two integration methods for integrating the weak form in meshfree methods [1].

In the first method, vertices of the background mesh are used as reference nodes and this is known as the element quadrature method. In order to successfully implement this technique, additional nodes need to be defined over the domain. Problems involving discontinuity of domain can be easily simulated with this method, as shown in Figure 1.5a.

The other integration method is known as cell quadrature. It requires a background grid of cells, which is independent of the domain. The spatial location of each Gauss point is checked in order to determine whether the Gauss point lies within or outside of the domain. Only the inside Gauss points are utilized for integration purposes. However, this method generates absurd results along curved boundaries, so is not widely used. However, this method provides satisfying results over regular domains.

## 1.8 DOMAIN OF INFLUENCE

The influence domain is an important parameter of meshfree methods, so its value must be chosen appropriately. The support domain size should be sufficiently large so as to prevent poor conditioning of matrices for the system. But increasing the size of the influence domain leads to a more computationally challenging task for making approximations and also in the assembly of the conductivity matrix. To obtain a good non-singular solution, the domain of influence is obtained by multiplying by a scaling parameter or factor ($d_{max}$). The range of scaling parameters [1] $d_{max}$ varies from problem to problem, and commonly lies in the range of $1 < d_{max} < 3$.

## 1.9 IMPOSITION OF BOUNDARY CONDITIONS

As the Kronecker delta property is not satisfied by this MLS approximation ($\Phi_I(x_J) \neq \delta_{IJ}$), the exact imposition of the essential boundary conditions (EBC) becomes a difficult task in EFGM, and a number of other methods have been suggested for imposing these essential boundary conditions. Belytschko [17] devised the Lagrange multiplier method. This method is quite efficient but it causes the system matrix to loses the positive definite and bandedness property. Krongauz and Belytschko [17, 18] established another technique, by blending with the finite element for the implementation of boundary conditions. Although this novel method reduces the complexities associated with the imposition of boundary conditions, numerical integration remains a difficult task. In the penalty method, the boundary conditions are applied easily, since it generates equations in a discrete form that are simple in form. Thus, obtaining the system matrix in this way forces it to remain positive and banded, although errors in selecting suitable penalty parameters may generate absurd solutions.

## 1.10 GOVERNING EQUATION

Let us consider a two-dimensional domain, being confined by small displacements in the domain $\Omega$, which is bounded by $\Gamma$. For this, the equation of equilibrium will be

$$\nabla \cdot \sigma + \mathbf{b} = 0 \text{ over } \Omega \tag{1.57}$$

where the stress tensor is represented by $\sigma$, the displacement field is represented by $u$, and $\mathbf{b}$ denotes the body force vector. The necessary boundary conditions are

$$\sigma + \mathbf{n} = \bar{t} \text{ over } \Gamma_t$$

$$\mathbf{u} = \bar{\mathbf{u}} \text{ over } \Gamma_u$$

where $\mathbf{n}$ is the outward unit normal to the domain $\Omega$.

The variational form of the equilibrium Eq. (1.57) can be expressed as

$$\int_{\Omega} \delta\varepsilon : \sigma \, d\Omega - \int_{\Omega} \delta\mathbf{u} \cdot \mathbf{b} \, d\Omega - \int_{\Gamma_t} \delta\mathbf{u} \cdot \bar{\mathbf{t}} \, d\Gamma - \delta\mathbf{W}_u\left(\mathbf{u}, \nabla_s \lambda\right) = 0 \tag{1.58}$$

where $\delta\varepsilon = \nabla_s(\delta\mathbf{u})$, $\nabla_s$ is a gradient operator, and $\delta\mathbf{W}_u$ is used to impose the essential boundary conditions. $\delta\mathbf{W}_u$ is essential, because, in relation to FEM (finite element method), since $\Phi_I\left(x_J\right) \neq \delta_{IJ}$, it is insufficient to impose the essential boundary conditions by the displacement of nodes. Different variations of $\mathbf{W}_u$ are possible. In this work, the Lagrange multiplier technique is used, so

$$\mathbf{W}_u(\mathbf{u}, \lambda) = \int_{\Gamma_u} \lambda \cdot (\mathbf{u} - \bar{\mathbf{u}}) d\Gamma \tag{1.59}$$

$$\delta\mathbf{W}_u(\mathbf{u}, \lambda) = \int_{\Gamma_u} \delta\lambda \cdot (\mathbf{u} - \bar{\mathbf{u}}) d\Gamma + \int_{\Gamma_u} \delta\mathbf{u} \cdot \lambda \, d\Gamma \tag{1.60}$$

Within the linear elastic regime, $\varepsilon = \nabla_s \mathbf{u}$ and $\boldsymbol{\sigma} = \mathbf{D}\boldsymbol{\varepsilon}$

where $\boldsymbol{\varepsilon}$ represents the strain vector and $\mathbf{D}$ is the constant material matrix. Let us take the discrete equation of the weak form, Eq. (1.58), having the boundary conditions in Eqs. (1.59) and (1.60). The Lagrange multiplier $\lambda$ will be given by

$$\lambda(\mathbf{x}) = N_I(s)\lambda_I, \qquad \mathbf{x} \in, u$$
$$\delta\lambda(\mathbf{x}) = N_I(s)\delta\lambda_I \qquad \mathbf{x} \in, u \tag{1.61}$$

where $N_{I(s)}$ is a Lagrange interpolant and $s$ denotes the length of the arc along the boundary. In Eq. (1.58), replacing $u(\mathbf{x})$ by the EFGM approximation, $u^h(\mathbf{x})$, and replacing the variation, $\delta u(\mathbf{x})$, by the EFGM approximation, $\delta u^h(\mathbf{x})$, $\delta u^h$ represents the variation of $u^h$.

$$u^h(\mathbf{x}) = \sum_{I=1}^{n} \Phi_I(\mathbf{x}) u_I$$

$$\delta u^h(\mathbf{x}) = \sum_{I=1}^{n} \Phi_I(\mathbf{x}) \delta u_I$$

The values of test function values $\delta u_I$ will be arbitrary, except over $\Gamma_u$, and thus can be eliminated from the equations. Substituting Eq. (1.61) into the weak form (1.58), gives:

$$\begin{bmatrix} \mathbf{K} & \mathbf{G} \\ \mathbf{G}^{\mathbf{T}} & 0 \end{bmatrix} \begin{Bmatrix} \mathbf{u} \\ \lambda \end{Bmatrix} = \begin{Bmatrix} \mathbf{f} \\ \mathbf{q} \end{Bmatrix} \tag{1.62}$$

where

$\mathbf{K}_{IJ} = \int_\Omega \mathbf{B}^T \mathbf{D} \mathbf{B}_J d\Omega$, $\mathbf{G}_{IK} = -\int_{T_u} \Phi_I \mathbf{N}_k d\Gamma_u$, $\mathbf{f}_I = \int_{\Gamma_t} \Phi_I \bar{\mathbf{t}}\, d\Gamma_t + \int_\Omega \Phi_I \mathbf{b}\, d\Omega$ and

$\mathbf{q}_k = -\int_\Gamma \mathbf{N_k} \bar{\mathbf{u}}\, d\Gamma_u$

with $\mathbf{B}_I = \begin{bmatrix} \Phi_{I,x} & 0 \\ 0 & \Phi_{I,y} \\ \Phi_{I,y} & \Phi_{I,x} \end{bmatrix}$, $\mathbf{N}_k = \begin{bmatrix} N_k & 0 \\ 0 & N_k \end{bmatrix}$, $\mathbf{D} = \frac{E}{1-v^2} \begin{bmatrix} 1 & v & 0 \\ v & 1 & 0 \\ 0 & 0 & \frac{1-v}{2} \end{bmatrix}$ for plane stress and $\mathbf{D} = \frac{E}{(1+v)(1-2v)} \begin{bmatrix} 1-v & v & 0 \\ v & 1-v & 0 \\ 0 & 0 & \frac{1-2v}{2} \end{bmatrix}$ for plane strain.

## 1.11 CRACK MODELING IN THE ELEMENT-FREE GALERKIN METHOD

In the modeling of crack tip fields, special criteria have been adopted for implementing singular functions related to elastic fracture; alternatively, more nodes can be added to the crack tip. The latter method can be challenging and tedious for problems of complex geometry. It was observed that simulation of the singular fields in a meshfree method is quite simple and trouble-free in comparison with FEM. Enrichment of a meshfree method can be done extrinsically or intrinsically.

### 1.11.1 Extrinsic MLS Enrichment

In extrinsic enrichment, trial functions are involved to include the first term of the near-tip asymptotic fields for the displacement. A closely associated function is blended with the polynomial expansion of the moving least square approximation. For example, the near-tip asymptotic field will be added to the solution for fracture mechanics problems. The approximation takes the form [14]

$$\mathbf{u}_{\alpha}^{h}(x) = \mathbf{c}^{T}(x)\mathbf{b}_{\alpha}(x) + \sum_{j=1}^{n_c} k_1^j Q_{1\alpha}^j(x) + k_2^j Q_{2\alpha}^j(x) \tag{1.63}$$

where $\mathbf{u}_{\alpha}^{h}(\mathbf{x})$ indicates the approximation for $u_{\alpha}(\mathbf{x})$, $\mathbf{c}(\mathbf{x})$ is a polynomial basis in the space coordinates, $n_c$ denotes the number of cracks in the model, $\mathbf{b}_{\alpha}(\mathbf{x})$ are the coefficients for the polynomial basis, and $k_1^j$ and $k_2^j$ are universal unknowns linked with crack $j$.

The coefficients $\mathbf{b}_{\alpha}(\mathbf{x})$ represent functions of the spatial coordinates that can be obtained by the MLS procedure. But extra terms come into the picture because of the involvement of the near-tip field, and hence the moving least square formulation is re-derived in order to satisfy the conditions for completeness. A weighted, discrete $L_2$ norm is written as

$$J = \sum_{i=1}^{n} w(\mathbf{x}-\mathbf{x}_I)\left[\mathbf{c}^{T}(\mathbf{x}_I)\mathbf{b}_{\alpha}(\mathbf{x}) + \sum_{j=1}^{n_c}\left[k_1^j Q_{1\alpha}^j(\mathbf{x}_I) + k_2^j Q_{2\alpha}^j(\mathbf{x}_I)\right] - u_{I\alpha}\right]^2 \tag{1.64}$$

where $n$ is the number of nodes in the vicinity of x that has a non-zero weight function $w(\mathbf{x} - \mathbf{x}_I)$, and $u_{I\alpha}$ denotes the component of the nodal value at $\mathbf{x}_I$. The consistency of $J$ with respect to $\mathbf{b}_{\alpha}(\mathbf{x})$ leads to

$$\mathbf{A}(\mathbf{x})\mathbf{b}_{\alpha}(\mathbf{x}) = \sum_{I=1}^{n} \mathbf{C}_I(\mathbf{x})\left\{u_{I\alpha} - \sum_{j=1}^{n_c}\left[k_1^j Q_{1\alpha}^j(\mathbf{x}_I) + k_2^j Q_{2\alpha}^j(\mathbf{x}_I)\right]\right\} \tag{1.65}$$

where

$$\mathbf{A}(\mathbf{x}) = \sum_{I=1}^{n} w(\mathbf{x}-\mathbf{x}_I)\mathbf{c}(\mathbf{x}_I)\mathbf{c}^{T}(\mathbf{x}_I)$$

$$\mathbf{C}_I(\mathbf{x}) = w(\mathbf{x}-\mathbf{x}_I)\mathbf{c}(\mathbf{x}_I)$$

Here $k_1^j$ and $k_2^j$ are the global parameters and they are assumed to be fixed in this method of calculating the parameter $\mathbf{b}_\alpha$. Solving Eq. (1.65) for $\mathbf{b}(\mathbf{x})$ gives

$$\left(\mathbf{b}_\alpha(\mathbf{x}) = \sum_{I=1}^{n} \mathbf{A}^{-1}(\mathbf{x})\mathbf{C}_I(\mathbf{x})\left\{u_{I\alpha} - \sum_{j=1}^{n_c}\left[k_1^j Q_{1\alpha}^j(\mathbf{x}_I) + k_2^j Q_{2\alpha}^j(\mathbf{x}_I)\right]\right\}\right. \tag{1.66}$$

Defining in terms $u_{I\alpha}$ (the nodal parameter) & the $k_1^j$ and $k_2^j$ (enriched field parameters):

$$\begin{aligned}\mathbf{u}_\alpha^h(\mathbf{x}) &= \sum_{I=1}^{n} \Phi_I(\mathbf{x})\left\{u_{I\alpha} - \sum_{j=1}^{n_c}\left[k_1^j Q_{1\alpha}^j(\mathbf{x}_I) + k_2^j Q_{2\alpha}^j(\mathbf{x}_I)\right]\right\} \\ &+ \sum_{j=1}^{n_c}\left[k_1^j Q_{1\alpha}^j(\mathbf{x}) + k_2^j Q_{2\alpha}^j(\mathbf{x})\right]\end{aligned} \tag{1.67}$$

$$\begin{aligned}\mathbf{u}_\alpha^h(\mathbf{x}) &= \sum_{I=1}^{n} \Phi_I(\mathbf{x})u_{I\alpha} + \sum_{j=1}^{n_c} k_1^j\left[Q_{1\alpha}^j(\mathbf{x}) - \sum_{I=1}^{n}\Phi_I(\mathbf{x})Q_{1\alpha}^j(\mathbf{x}_I)\right] \\ &+ \sum_{j=1}^{n_c} k_2^j\left[Q_{2\alpha}^j(\mathbf{x}) - \sum_{I=1}^{n}\Phi_I(\mathbf{x})Q_{2\alpha}^j(\mathbf{x}_I)\right]\end{aligned} \tag{1.68}$$

The value of shape function is given as

$$\Phi_I(\mathbf{x}) = \mathbf{c}^T(\mathbf{x})\mathbf{A}^{-1}(\mathbf{x})\mathbf{C}_I(\mathbf{x}) \tag{1.69}$$

Eq. (1.68) will be written as

$$\mathbf{u}_\alpha^h(\mathbf{x}) = \sum_{I=1}^{n} \Phi_I(\mathbf{x})\tilde{u}_{I\alpha} + \sum_{j=1}^{n_c}\left[k_1^j Q_{1\alpha}^j(\mathbf{x}) + k_2^j Q_{2\alpha}^j(\mathbf{x})\right] \tag{1.70}$$

where the modified nodal coefficients, $\tilde{u}_{I\alpha}$, are

$$\tilde{u}_{I\alpha} = u_{I\alpha} - \sum_{j=1}^{n_c}\left[k_1^j\, Q_{1\alpha}^j(\mathbf{x}_I) + k_2^j\, Q_{2\alpha}^j(\mathbf{x}_I)\right] \tag{1.71}$$

### 1.11.2 Intrinsic MLS Enrichment

The approximate solution can also be intrinsically enriched [14] by modifying the basis with a special function. For example, in fracture mechanics, this is achieved by including the asymptotic near-tip displacement field, or $\sqrt{r}$, as a crucial parameter.

The accuracy of the solution will depend upon the choice of functions. For greater accuracy, we use the full asymptotic field, but, for higher computational speed, the basis includes only the $\sqrt{r}$ function. Both these methods will be explained in the subsequent sections. The whole of the near-tip displacement field is included in the basis case of full intrinsic enrichment. With some trigonometric calculations, it can be shown [1] that all the functions in Eq. (1.18) can be represented by the following basis:

$$\mathbf{P}^T(\mathbf{x}) = \left[1, x, y, \sqrt{r}\cos\frac{\theta}{2}, \sqrt{r}\sin\frac{\theta}{2}, \sqrt{r}\sin\frac{\theta}{2}\sin\theta, \sqrt{r}\cos\frac{\theta}{2}\sin\theta\right] \tag{1.72}$$

This basis generates the approximations as follows:

$$u^h(\mathbf{x}) = \sum_{i=1}^{n} \mathbf{c}^T(\mathbf{x})\mathbf{A}^{-1}(\mathbf{x})\mathbf{C}_I(\mathbf{x})u_I \tag{1.73}$$

where $\mathbf{\Phi}_I(\mathbf{x})$ is the EFGM shape function (enriched).

As compared with the extrinsic techniques, this method requires no additional unknowns, though, due to the enhanced size of the basis, extra computational time is needed for inversion of the moment matrix, $\mathbf{A}(\mathbf{x})$. The basis of each crack needs some extra terms to be added in the case of multiple cracks.

## 1.12 INTEGRATION INTEGRAL

The interaction integral forms an efficient method for determining the parameters of a fracture under mixed-mode conditions. The expression for the path-independent $J$-integral [14] for a domain with an existing crack is given as:

$$J = \int_\Gamma \left( W\delta_{1j} - \sigma_{ij}\frac{\partial u_i}{\partial x_1} \right) n_j \, d\Gamma \tag{1.74}$$

where $W = \int \sigma_{ij}\, d\varepsilon_{ij}$ denotes strain energy density and $n_j$ represents the component of outward normal unit vector normal over an arbitrary path $\Gamma$, enclosing the region near the crack tip. The strain energy for linear elastic materials is given as $W = \sigma_{ij}\varepsilon_{ij}/2$.

The path-independent property enables the $J$-integral to be calculated with far field information, which is more accurate than the near-tip solution. To improve its utility, the contour integral in Eq. (1.74) is modified to the equivalent domain form by the application of the divergence theorem.

$$\begin{aligned} J = &\int_A \left( \sigma_{ij}\frac{\partial u_i}{\partial x_1} - W\delta_{1j} \right) \frac{\partial q}{\partial x_j} dA \\ &+ \int_A \frac{\partial}{\partial x_j}\left( \sigma_{ij}\frac{\partial u_i}{\partial x_1} - W\delta_{1j} \right) q dA \end{aligned} \tag{1.75}$$

where $A$ denotes the area within the interior of contour and $q$ is a linearly varying weight function, having a value of *unity* at the crack tip, and *zero* along the domain boundary, and being arbitrary at rest points within the domain. By expanding the second integrand, Eq. (1.75) reduces to

$$J=\int_A\left(\sigma_{ij}\frac{\partial u_i}{\partial x_1}-W\delta_{1j}\right)\frac{\partial q}{\partial x_j}dA$$
$$+\int_A\left(\frac{\partial\sigma_{ij}}{\partial x_j}\frac{\partial u_i}{\partial x_1}+\sigma_{ij}\frac{\partial^2 u_i}{\partial x_j\partial x_1}-\sigma_{ij}\frac{\partial\varepsilon_{ij}}{\partial x_1}\right)qdA \tag{1.76}$$

Using equilibrium $\left(\partial\sigma_{ij}/\partial x_j=0\right)$ and compatibility $\varepsilon_{ij}=\partial u_i/\partial x_j$ Eq. (1.76) gives us the following:

$$J=\int_A\left(\sigma_{ij}\frac{\partial u_i}{\partial x_1}-W\delta_{1j}\right)\frac{\partial q}{\partial x_j}dA \tag{1.77}$$

This represents the conventional $J$-integral for homogeneous space.

Now, let us consider a cracked body in two equilibrium states. State 1 represents the actual state, whereas state 2 is defined to be an auxiliary one. Another equilibrium state (state $S$) can be obtained by the superimposition of these two states:

$$J^{(S)}=\int_A\left(\left(\sigma_{ij}^{(1)}+\sigma_{ij}^{(2)}\right)\frac{\partial\left(u_i^{(1)}+u_i^{(2)}\right)}{\partial x_1}-W^{(S)}\delta_{1j}\right)\frac{\partial q}{\partial x_j}dA \tag{1.78}$$

where superscript $i=1$, 2, and $S$ indicates fields and quantities associated with state $i$. Moreover, the strain energy density for the superimposed state is given as:

$$W^{(S)}=\frac{1}{2}\left(\sigma_{ij}^{(1)}+\sigma_{ij}^{(2)}\right)\left(\varepsilon_{ij}^{(1)}+\varepsilon_{ij}^{(2)}\right) \tag{1.79}$$

By expanding Eq. (1.78),

$$J^{(S)}=J^{(2)}+J^{(1)}+M^{(1,2)} \tag{1.80}$$

where

$$J^{(1)}=\int_A\left(\sigma_{ij}^{(1)}\frac{\partial u_i^{(1)}}{\partial x_1}-W^{(1)}\delta_{1j}\right)\frac{\partial q}{\partial x_j}dA \tag{1.81}$$

and

$$J^{(2)}=\int_A\left(\sigma_{ij}^{(2)}\frac{\partial u_i^{(2)}}{\partial x_1}-W^{(2)}\delta_{1j}\right)\frac{\partial q}{\partial x_j}dA \tag{1.82}$$

are the $J$-integral for states 1 and 2, respectively, and

$$M^{(1,2)} = \int_A \left( \sigma_{ij}^{(1)} \frac{\partial u_i^{(2)}}{\partial x_1} + \sigma_{ij}^{(2)} \frac{\partial u_i^{(1)}}{\partial x_1} - W^{(1,2)} \delta_{1j} \right) \frac{\partial q}{\partial x_j} dA \tag{1.83}$$

is the expression for the interaction integral.

In Eqs. (1.26–1.28), $W^{(1)} = \frac{1}{2}\sigma_{ij}^{(1)}\varepsilon_{ij}^{(1)}$, $W^{(2)} = \frac{1}{2}\sigma_{ij}^{(2)}\varepsilon_{ij}^{(2)}$, and $W^{(1,2)} = \frac{1}{2}\left(\sigma_{ij}^{(1)}\varepsilon_{ij}^{(1)} + \sigma_{ij}^{(2)}\varepsilon_{ij}^{(2)}\right)$ indicate strain energy densities, which must also satisfy

$$W^{(S)} = W^{(1)} + W^{(2)} + W^{(1,2)} \tag{1.84}$$

For linear elastic material under mixed-mode loading conditions, the $J$-integral can also be equated to the energy release rate, so that the $J$-integral can be written as:

$$J = \frac{1}{E^*}\left(K_I^2 + K_{II}^2\right) \tag{1.85}$$

where

$$E^* = \begin{cases} \dfrac{E}{1-\nu^2} & \text{for plane strain} \\ E & \text{for plane stress} \end{cases}$$

Applying Eq. (1.85) to states 1, 2, and state $S$ gives us

$$J^{(1)} = \frac{1}{E^*}\left(K_I^{(1)^2} + K_{II}^{(1)^2}\right) \tag{1.86}$$

$$J^{(2)} = \frac{1}{E^*}\left(K_I^{(2)^2} + K_{II}^{(2)^2}\right) \tag{1.87}$$

and

$$\begin{aligned} J^{(S)} &= \frac{1}{E^*}\left[\left(K_I^{(1)} + K_I^{(2)}\right)^2 + \left(K_{II}^{(1)} + K_{II}^{(2)}\right)^2\right] \\ &= \frac{1}{E^*}\left[\left(K_I^{(1)^2} + K_{II}^{(1)^2}\right) + \left(K_I^{(1)^2} + K_{II}^{(2)^2}\right) + 2\left(K_I^{(1)}K_I^{(2)} + K_{II}^{(1)}K_{II}^{(2)}\right)\right] \\ &= J^{(I)} + J^{(2)} + \frac{2}{E^*}\left(K_I^{(1)}K_I^{(2)} + K_{II}^{(1)}K_{II}^{(2)}\right) \end{aligned} \tag{1.88}$$

Comparing Eqs (1.80) and (1.88),

$$M^{(1,2)} = \frac{2}{E^*}\left[\left(K_I^{(1)}K_I^{(2)} + K_{II}^{(1)}K_{II}^{(2)}\right)\right] \tag{1.89}$$

The independent stress intensity factors for the defined problem can be calculated by selecting the auxiliary state appropriately. For example, if state 2 is chosen to be state 1, then $K_I^{(2)} = 1$ and $K_{II}^{(2)} = 0$. Hence, Eq. (1.89) can be reduced to

$$M^{(1,I)} = \frac{2K_I^{(1)}}{E^*} \tag{1.90}$$

from which

$$K_I^{(1)} = \frac{M^{(1,I)}E^*}{2} \tag{1.91}$$

Similarly, if state 2 is considered to be state II, then $K_I^{(2)} = 0$ and $K_{II}^{(2)} = 1$. Thus, with similar considerations,

$$K_{II}^{(1)} = \frac{M^{(1,II)}E^*}{2} \tag{1.92}$$

In this way, numerical computation of interaction integral enables us to predict the mixed-mode stress intensity factors.

## 1.13 APPLICATIONS OF ELEMENT-FREE GALERKIN METHODS TO COMPUTATIONAL FRACTURE MECHANICS

Several numerical problems pertaining to the area of computational fracture mechanics are discussed below, to establish the capability and robustness of the element-free Galerkin method. The problems discussed below are part of the author's research work.

### 1.13.1 Crack Modeling under Mechanical Loads

For this simulation, a domain with a crack is considered with dimensions of $H$=2 cm, $W$=1 cm, and crack length=40 mm, as shown in Figure 1.6. ASTM 36 steel, having an elastic modulus ($E$)=200 GPa and Poisson's ratio ($\nu$)=0.3, is considered. The applied stress ($\sigma_o$) is=100 MPa,

The variation in stress intensity factors by changing crack angle inclination have been simulated. The spatial location of the crack is at a distance of 100 mm from the bottom and $W$/2 from the edge of the domain. Mode-I mechanical loading has been employed for the present simulation.

The problem geometry is modeled using 800 nodes. Six-point Gauss quadrature is utilized to carry out the numerical integration. The simulation has been carried out under plane stress state. Stress intensity factors for both modes (1 and 2) were computed using the M-integral technique.

The crack length considered is $2a$=40 mm. Figure 1.7a represents a single edge cracked geometry subjected to mechanical loads. The stress intensity factors have

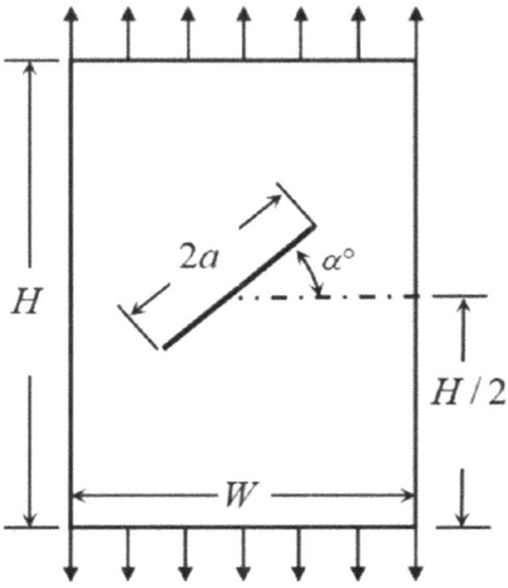


**FIGURE 1.6** Problem geometries and their dimensions along with boundary conditions.

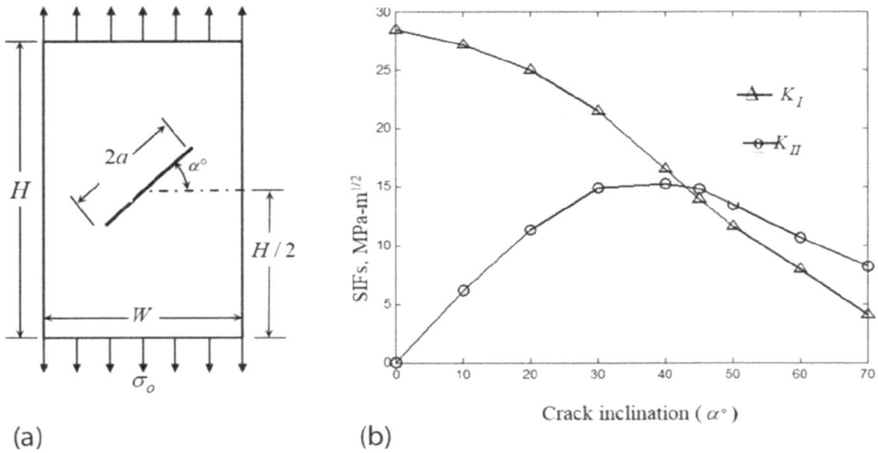


**FIGURE 1.7** Effect of crack inclination (α) on $K_I$ and $K_{II}$ of a center crack.

been evaluated for different values of crack angle ($\alpha$) shown in Figure 1.7b. A maximum value of mode-1 stress intensity factor is achieved as 28.5 MPA$\sqrt{\text{m}}$ at $\alpha=0°$ (close to the exact analytical value [18] for $\alpha=0°$, which is 28.7 MPA$\sqrt{\text{m}}$), whereas a maximum value of mode-2 stress intensity factor, i.e. $K_{II}$, is calculated for $\alpha=40°$. From the results shown in Figure 1.7b, it can be seen that, with an increase in $\alpha$, the value of $K_I$ falls continuously, whereas $K_{II}$, after attaining its maximum value at $\alpha=40°$, starts to decrease, as shown in Figure 1.7b. At $\alpha=43°$, the values of $K_I$ and $K_{II}$ become almost equal, which can be predicted from the meeting of the two curves in Figure 1.7b. In order to clearly visualize the stress distribution around the crack,

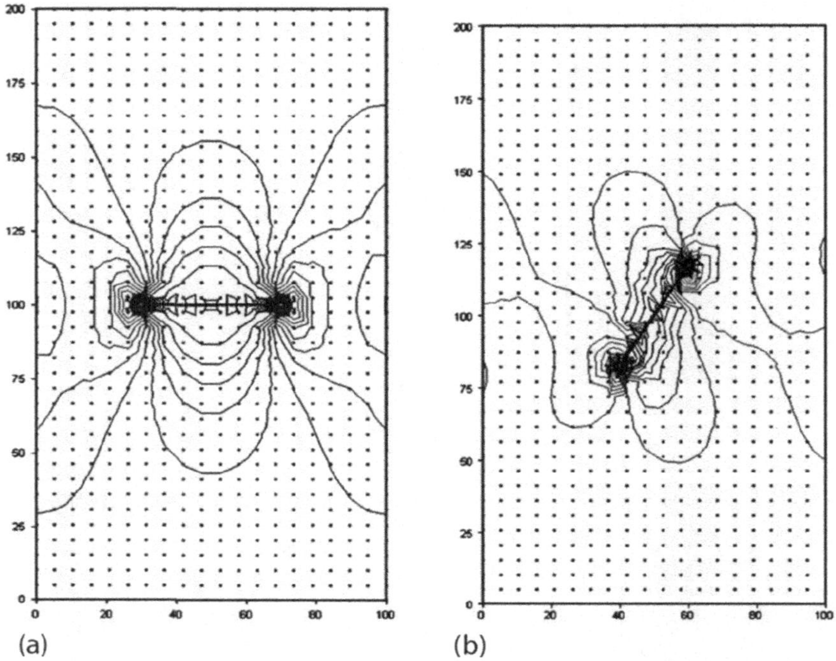


**FIGURE 1.8** Contour plots of stress ($\sigma_{yy}$) for different inclinations subjected to mechanical loading.

the contour plots for components have been generated for the problem domain. An inclination of 0° and 60° have been used to model and simulate inclined crack geometry, as shown in Figures 1.8a and Figure 1.8b.

### 1.13.2 Modeling of Vertical Bi-Material Interface

A prismatic bar of dimensions $L \times D$ is modeled, imposed with a force at the free end as shown in Figure 1.9. This problem was modeled with the state of plane stress having an elastic modulus as $E_1 = 20 \times 10^{11}$ unit, $\nu_1 = 0.3$, $E_2 = 0.2 \times 10^{11}$ unit, and $\nu_2 = 0.3$, and the geometrical data are D = 1 unit and L = 4 unit. The shear force is P = 10000 unit. The solutions were calculated with a linear basis function along with cubic spline weight and a $d_{max} = 1.1$.

The distribution of strains $\varepsilon_{xx}$ and $\varepsilon_{yy}$ are shown in Figure 1.10 over the length of the beam at the top and bottom locations, respectively.

### 1.13.3 Modelling of Bi-Metallic Interfacial Edge Crack

A bi-metallic plate with a crack length $a$ was subjected to a tensile load at the top and bottom edges as shown in Figure 1.11. The geometrical dimensions are

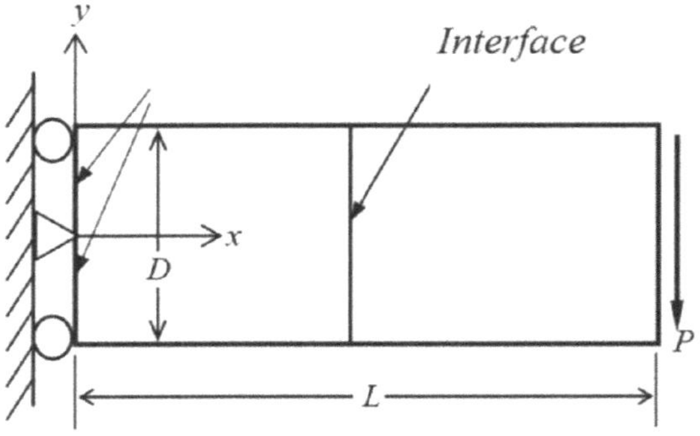


FIGURE 1.9 A two-dimensional bi-material beam with traction.

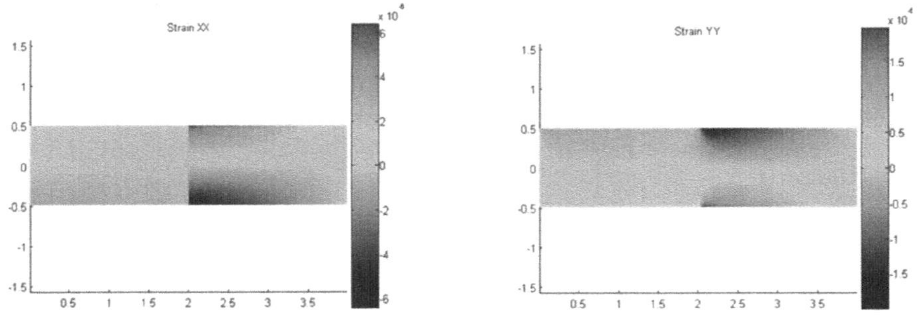


FIGURE 1.10 Distribution of strains over the domain.

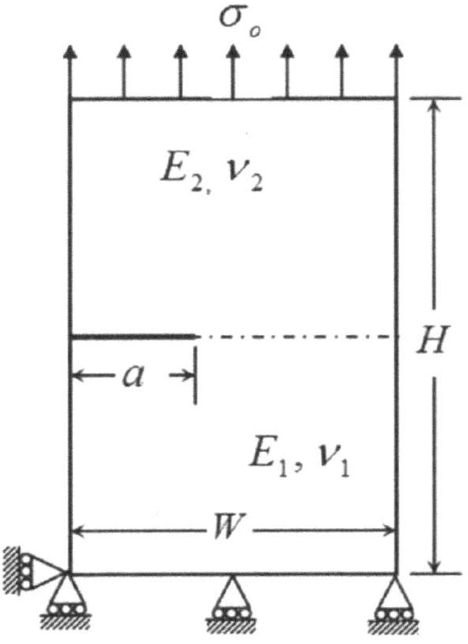


FIGURE 1.11 A bi-metallic interface edge crack.

$W=3$ units, $H=9$ units, and applied stress $\sigma_o=1$ unit. For both sides, Poisson's ratio is $\nu_1 = \nu_2 = 0.3$. Simulations were being carried out for varying material modulus ratios, so that the value of the proposed criteria is established. In order to present the validity of the proposed criteria for interface crack problems, the results are calculated for several different ratios of Young's modulus, $E_2/E_1=2$, 3, 10, 100, with $E_1$ as constant at 100 units. Also, the values of $a/W=0.2$, 0.3, 0.4, 0.5, 0.6 are taken into account. The normalized stress intensity factors are shown in Figures 1.12–1.15, with the results marked as Ref. [19] and Ref. [20], respectively.

Figure 1.12 represents the trend of normalized stress intensity factors (SIFs), i.e., $K_I$ and $K_{II}$, by varying the length of the crack, having $E_2/E_1=2$. The plots represent that for an increase in crack length, and $K_I$ also increases whereas $K_{II}$ decreases. Next, the EFGM results are calculated for $E_2/E_1=3$, 10, and 100 as shown in Figures 1.13–1.15, respectively. It can be clearly seen from Figures 1.12–1.15 that numerical results computed using the element-free Galerkin method closely match the reference values, having a standard deviation of less than 5%.

Furthermore, to clearly visualize the stress fields, the contour plots of stress and strain components were generated using varying $E_2/E_1$ ratios, with $a/W=0.4$. Figure 1.16a and Figure 1.16b represent the stress and strain contours, respectively, for $E_2/E_1=1$, i.e., contour plots reveal the continuity and symmetry of stress–strain about the $x$-axis.

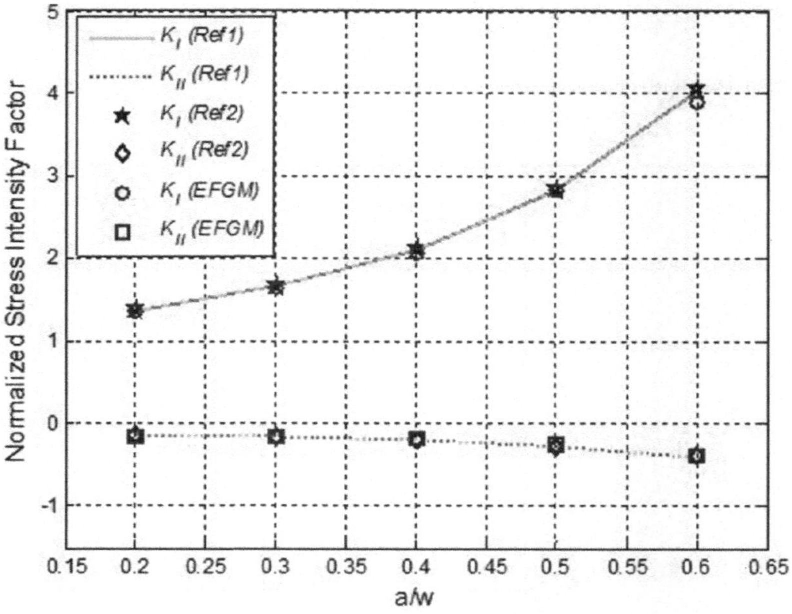


**FIGURE 1.12** SIF variation for $E_2/E_1=2$.

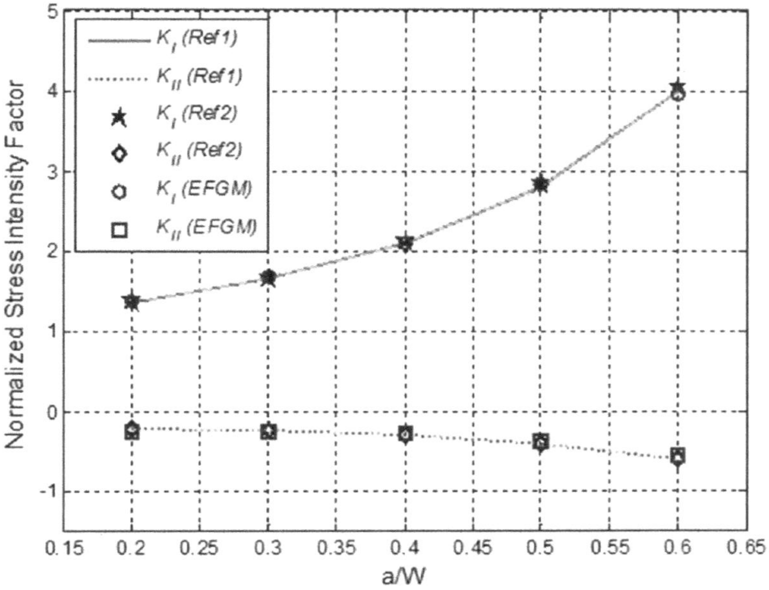


**FIGURE 1.13** SIF variation for $E_2/E_1 = 3$.

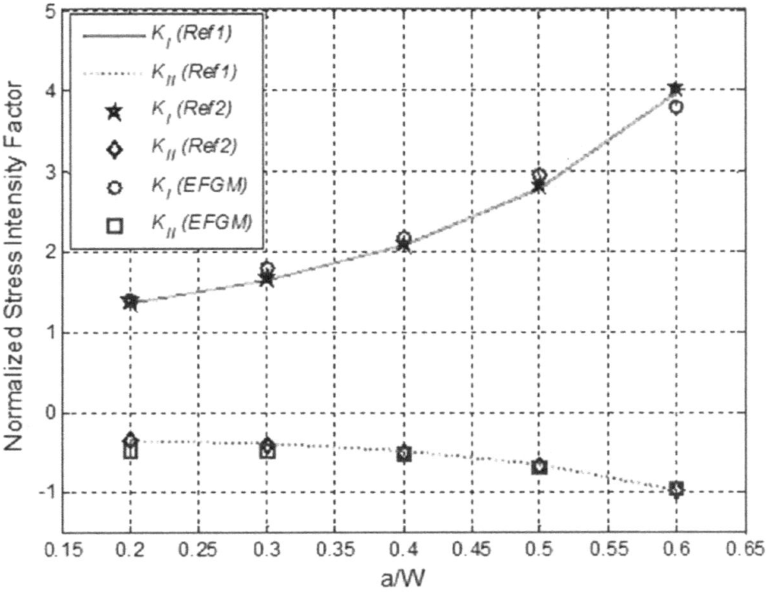


**FIGURE 1.14** SIF variation for $E_2/E_1 = 10$.

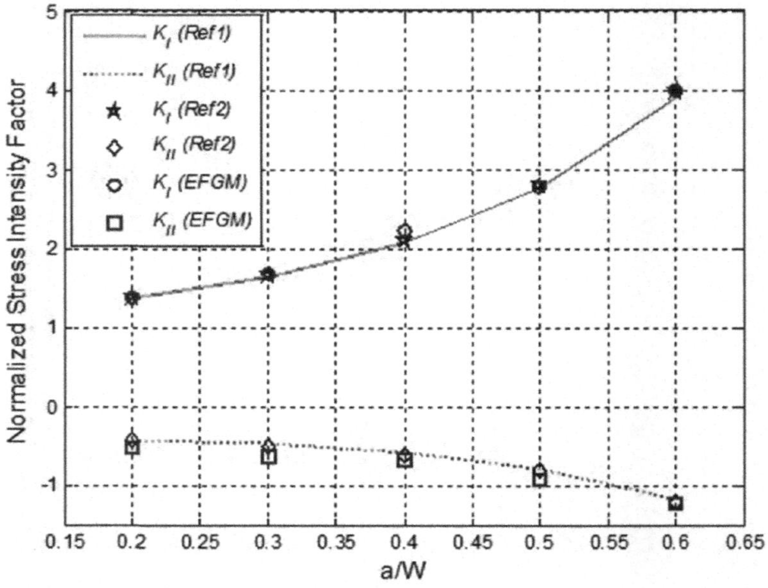


**FIGURE 1.15** SIF variation for $E_2/E_1 = 100$.

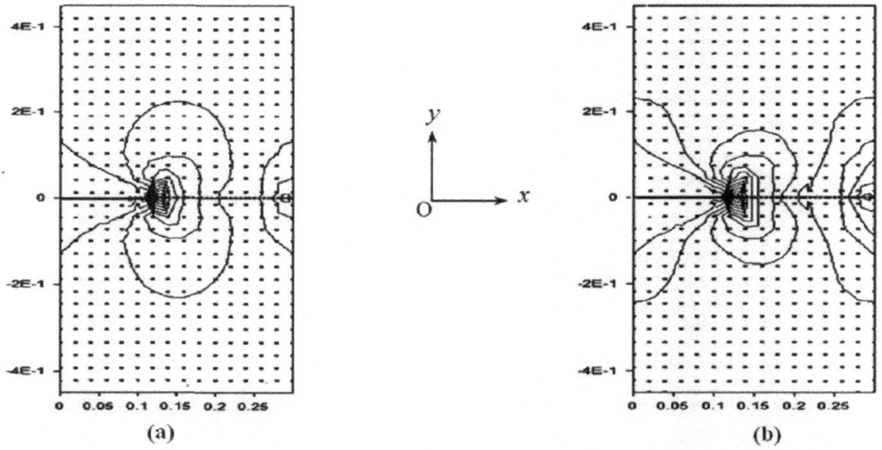


**FIGURE 1.16** $\sigma_{xx}$ and $\varepsilon_{xx}$ contours for $E_2/E_1 = 1$.

Similar types of contours have also been generated for varying ratios of $E_2/E_1 = 2$, 10, and 100. For $E_2/E_1 = 2$, contours of $\sigma_{YY}$ are symmetric but exhibit a minute deviation as observed in Figure 1.17a, but the strain fields exhibit a small discontinuity at the interface. Furthermore, the stress plots were generated for $E_2/E_1 = 10$. A small deviation of $\sigma_{YY}$ from the previous case is obtained, as displayed in Figure 1.18a, but

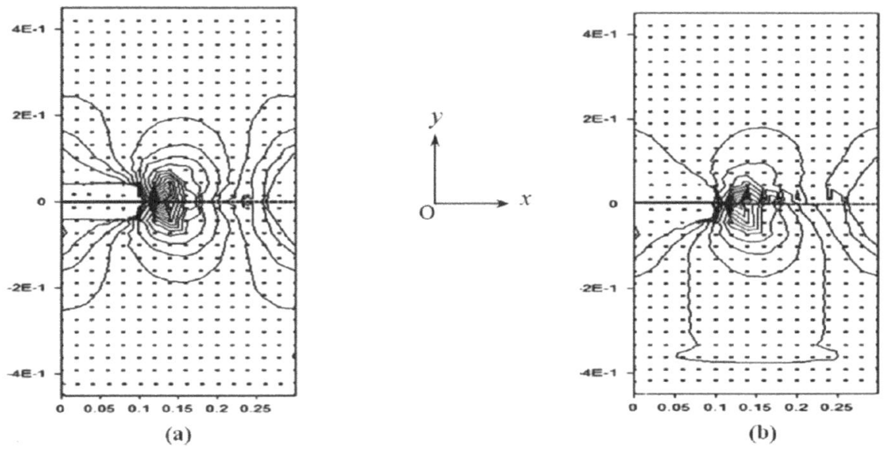


**FIGURE 1.17** $\sigma_{xx}$ and $\varepsilon_{xx}$ contours for $E_2/E_1=2$.

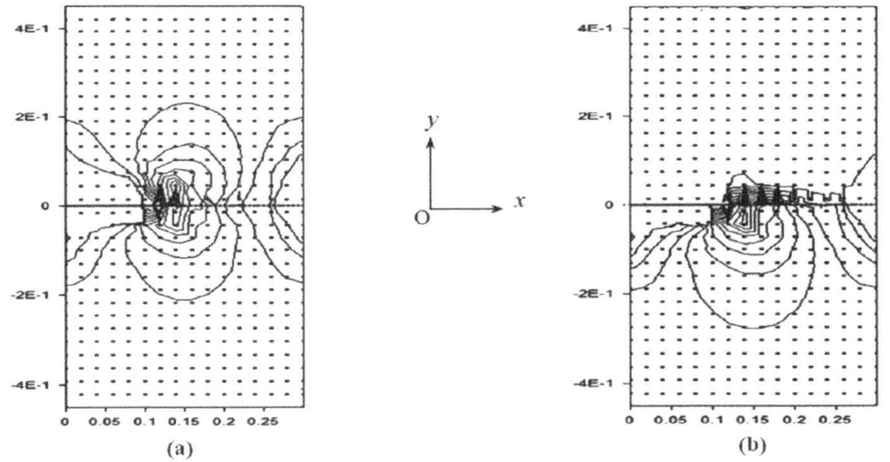


**FIGURE 1.18** $\sigma_{xx}$ and $\varepsilon_{xx}$ contours for $E_2/E_1=10$.

this higher modulus ratio generates a strongly discontinuous strain field, as shown in Figure 1.18b. A similar type of field plot was also generated for a higher material modulus ratio of $E_2/E_1=100$. Again, we obtained a nearly continuous and symmetrical stress field for $\sigma_{YY}$, as shown in Figures 1.19a and 1.20a, whereas the strain fields are significantly discontinuous, as can be seen in Figures 1.19b and 1.20b. All these simulations reveal that the presence of the interface and the value of the material modulus ratio play an important role in deciding the magnitude of the discontinuity of the strain field.

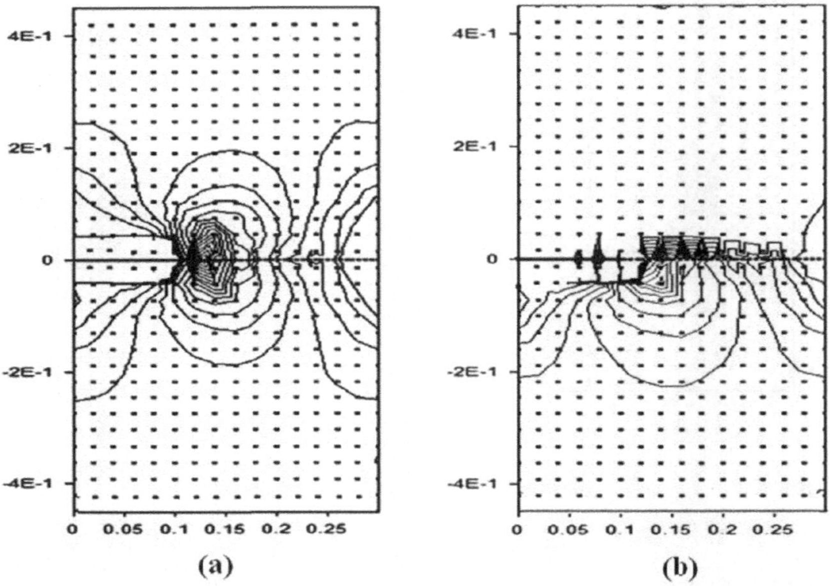


**FIGURE 1.19** $\sigma_{xx}$ and $\varepsilon_{xx}$ contours for $E_2/E_1 = 100$.

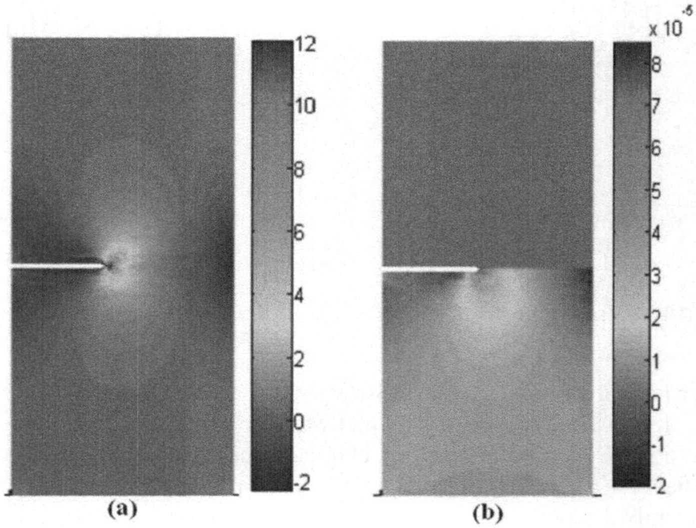


**FIGURE 1.20** $\sigma_{xx}$ and $\varepsilon_{xx}$ contours for $E_2/E_1 = 100$.

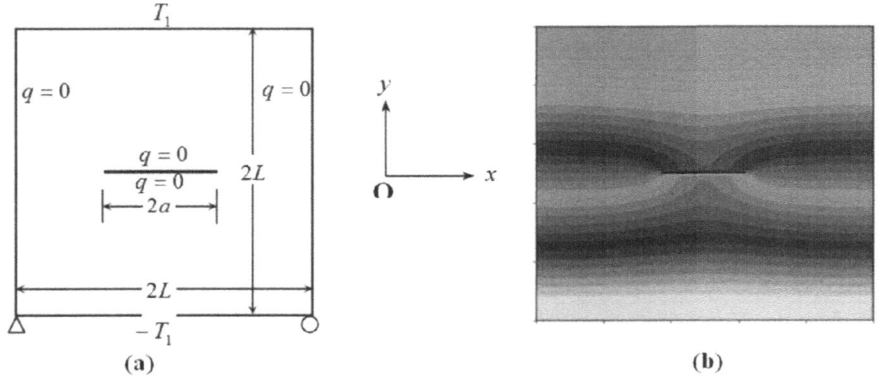


**FIGURE 1.21** Adiabatic center crack: (a) Problem geometry, (b) temperature profile.

### 1.13.4 Modeling of Thermoelastic Fracture

#### 1.13.4.1 Centre Crack in Square Domain

Next, let us take a square geometry with a center with two varying sets of thermal boundaries:

- An adiabatic boundary condition with a heat flux normal to the crack surface;
- An isothermal boundary condition where the crack is subjected to different temperatures.

Figure 1.21 represents a square domain subjected to both mechanical and thermal boundary values. For this, the length is considered as $L=1$ with $a=0.1$ to 0.6, increasing by a value of 0.1 units. Temperatures are imposed at the upper and lower boundaries of the domain.

Also, the right and left faces are assumed to be insulated. The crack is modeled as an adiabatic boundary and thus a discontinuous temperature field is generated along them. Figure 1.21b presents the contour of temperature fields within the domain. The location of strong discontinuity creates a discontinuous temperature field as simulated.

Furthermore, the same problem geometry is modeled, using a new constraint for the thermal boundary. Figure 1.22a represents a domain with a strong central discontinuity crack imposed to the isothermal boundary conditions at all four extremities. Now, the crack line is assumed to be a part of the essential boundary and imposed with some defined value of temperature. This change of boundary condition leads to the generation of a continuous temperature field through the crack boundary, as shown in Figure 1.22b. For the first case, a pure sliding mode (mode-II) is generated, whereas imposing the second set of boundary conditions generates a pure opening

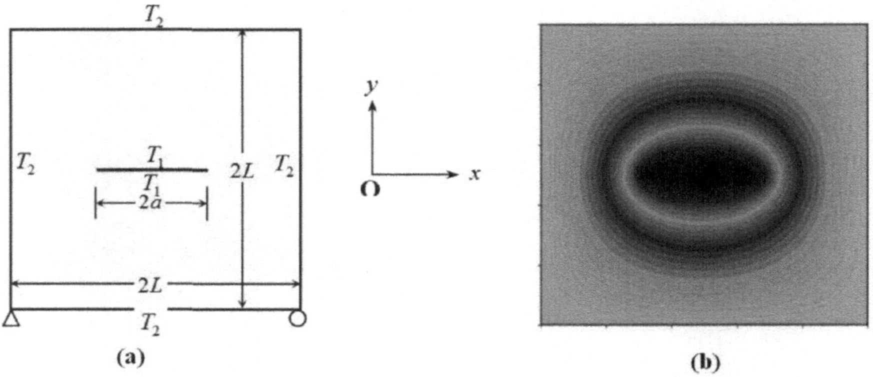


**FIGURE 1.22** Isothermal center crack: (a) Problem geometry, (b) temperature profile.

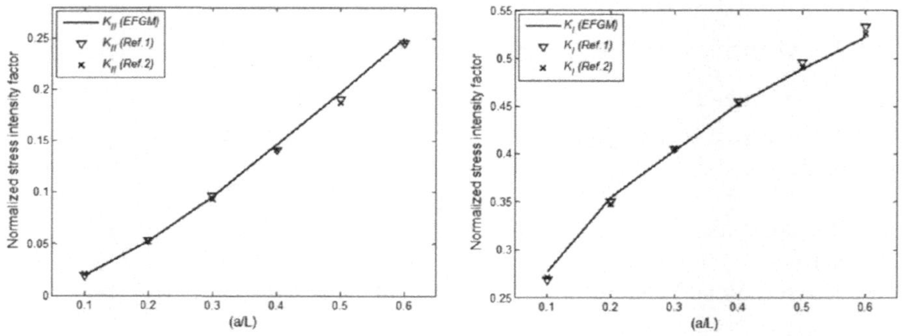


**FIGURE 1.23** Variation in a normalized stress intensity factor with crack length.

mode (mode-I). The values of stress intensity factor ($K_{II}$) have been calculated for the varying $a/L$ ratio, with 1152 nodes over the entire domain. The value of SIFs is presented in normalized form. Figure 1.23a represents the variation of the normalized mode-II stress intensity factor for an adiabatic boundary condition, while Figure 1.23b shows the trend of the normalized mode-I stress intensity factor for an isothermal boundary condition for a different $a/L$ ratio. The numerical results calculated by EFGM are found to be in close agreement with the reference values [21, 22].

### 1.13.5 Thermal Fracture in Coatings

In coatings, the component has a blend of discontinuities. The interface forms the weak part and any crack included forms a strong discontinuity. Intrinsic enrichment has been utilized to model the crack along with jump function criteria to model the weakly discontinuous interface part.

A uniform nodal density of ($25 \times 37$) nodes has been defined for problem simulation. Thermal loading conditions have been applied to the problem domain, with

**TABLE 1.1**
**Materials and Their Mechanical Properties**

| S. No. | Material | Young's modulus | Poisson's ratio | Coefficient of thermal expansion |
|---|---|---|---|---|
| | Steel | 210 GPa | 0.3 | $12\times10^{-6}$ m/m K |
| | Zinc | 100 GPa | 0.3 | $29\times10^{-6}$ m/m K |

height as 3 units and width as 2 units, along with a temperature gradient of 100°C. A plane stress condition is assumed to prevail over the domain. The crack surface is not subjected to any specific boundary conditions. The interaction between stresses and strains developed due to thermal loading with strong discontinuity is explored. A layer thickness of 0.5 units is considered for each layer. The primary crack is situated at a spatial location of $x=0$ and $y=B/2$, with an inclination of $\alpha=0°$. The present study explores the effect of substrate rigidity and variation in the material properties on cracked specimen. A maximum variation of 3–5% was found between the finite element method and the element-free Galerkin method. Table 1.1 shows the mechanical properties of the materials used for simulation. A lower value of error suggests that the present EFGM model is robust and has a high simulation efficiency. Table 1.2 represents the results of a comparative simulation obtained, using EFGM and FEM.

### 1.13.5.1 Edge Crack with a Thermal Load

A geometry having similar dimensions is considered for this simulation, as shown in Figure 1.24. The results for stress intensity factors obtained for varying inclination angles are plotted in Figure 1.25a–b. It is seen that the value of the mode-1 stress intensity factor is inversely proportional to the crank orientation angle. The result worth mentioning here is that the coefficient of thermal expansion of zinc or steel plays an important role in determining the SIFs of unimaterial geometry, as shown in Table 1.2.

The values in Table 1.2 show that a homogeneous plate of steel has the lowest value of mode-I SIF because of the low thermal expansion coefficient, thereby causing less expansion than the other configurations. The contour plots of stress in the '$y$' ($\sigma_{yy}$) direction for all the configurations are generated and are shown in Figure 1.26a–c.

## 1.14 CONCLUSION

This chapter presents a brief description of various meshfree methods, along with their developments over the course of time. A brief description of one of the most widely used and robust meshfree methods, known as the element-free Galerkin method, is explored. Theoretical aspects of the element-free Galerkin method, including the terminology and basic concepts, have been introduced. Special focus

**TABLE 1.2**
**Values of Mixed-Mode SIFs for All Configurations**

| Single edge crack (Crack length $a = 0.2$) | | SIFs for steel/zinc/steel configuration | | SIFs for zinc configuration | | SIFs for zinc/steel/zinc configuration | | SIFs for steel configuration | |
|---|---|---|---|---|---|---|---|---|---|
| Crack inclination angle (degree) | | $K_I$ (MPa-m$^{1/2}$) | $K_{II}$ (MPa-m$^{1/2}$) | $K_I$ (MPa-m$^{1/2}$) | $K_{II}$ (MPa-m$^{1/2}$) | $K_I$ (MPa-m$^{1/2}$) | $K_{II}$ (MPa-m$^{1/2}$) | $K_I$ (MPa-m$^{1/2}$) | $K_{II}$ (MPa-m$^{1/2}$) |
| 0 | EFG | 2.0049 | 0.0185 | 2.2140 | 0.0179 | 2.1402 | 0.0154 | 1.9238 | 0.0156 |
| | FEM | 2.0505 | 0.0179 | 2.2445 | 0.0172 | 2.0620 | 0.0179 | 1.8493 | 0.0150 |
| 10 | EFG | 1.7234 | 0.1863 | 1.9134 | 0.2884 | 1.8574 | 0.2032 | 1.6627 | 0.1811 |
| | FEM | 1.6698 | 0.1807 | 1.9186 | 0.2789 | 1.9251 | 0.1978 | 1.5983 | 0.1941 |
| 20 | EFG | 1.9699 | 0.4385 | 2.1814 | 0.4888 | 2.1147 | 0.4759 | 1.8956 | 0.4247 |
| | FEM | 1.9086 | 0.4209 | 2.1057 | 0.4742 | 2.0459 | 0.4911 | 1.8339 | 0.4382 |
| 30 | EFG | 1.7632 | 0.5120 | 1.9564 | 0.5707 | 1.9007 | 0.5562 | 1.7001 | 0.4959 |
| | FEM | 1.7979 | 0.4909 | 2.0387 | 0.5827 | 1.8332 | 0.5411 | 1.6623 | 0.4810 |
| 40 | EFG | 1.4324 | 0.6113 | 1.6019 | 0.6850 | 1.5657 | 0.6705 | 1.3920 | 0.5952 |
| | FEM | 1.3894 | 0.6296 | 1.5424 | 0.6596 | 1.5085 | 0.6429 | 1.4506 | 0.5702 |
| 50 | EFG | 1.1147 | 0.7747 | 1.2510 | 0.8873 | 1.2270 | 0.8660 | 1.0871 | 0.7624 |
| | FEM | 1.1366 | 0.7427 | 1.3036 | 0.9060 | 1.1834 | 0.8425 | 1.0629 | 0.7396 |
| 60 | EFG | 0.9010 | 0.7718 | 1.1195 | 0.8167 | 1.0063 | 0.8083 | 0.8859 | 0.7097 |
| | FEM | 0.8729 | 0.7486 | 1.0802 | 0.7899 | 1.0430 | 0.7869 | 0.8516 | 0.6823 |

SIF: Stress intensity factor, EFG: Element free Galerkin, FEM: Finite Element Method.

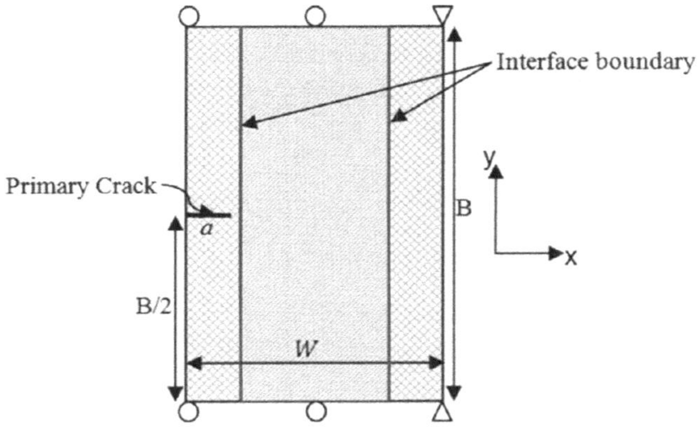


**FIGURE 1.24** Problem geometry along with boundary conditions.

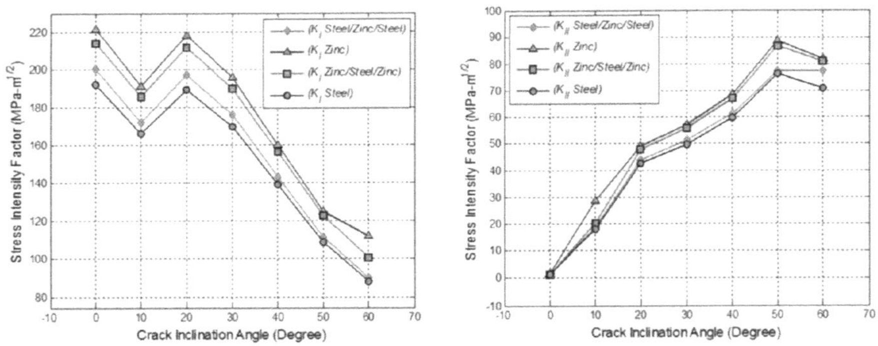


(a) (b)

**FIGURE 1.25** Plots of SIFs (a) mode-I, (b) mode-II.

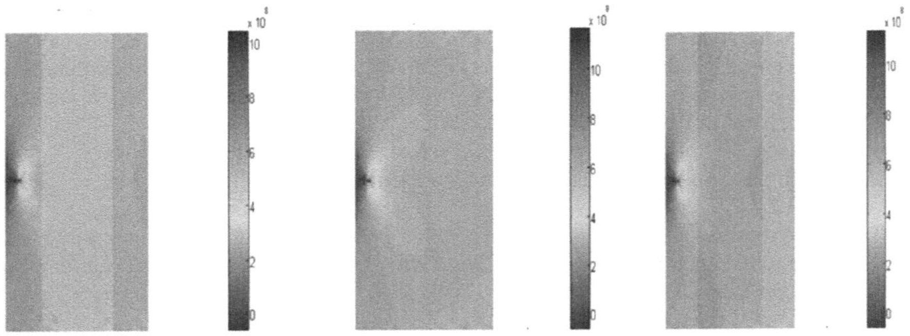

(a) (b) (c)

**FIGURE 1.26** Contour plot of $\sigma_{yy}$ (a) steel/zinc/steel configuration, (b) mono-material configuration, and (c) zinc/steel/zinc configuration.

has been applied toward the use of the element-free Galerkin technique in modeling and simulation of problems of fracture mechanics. A number of numerical problems have been discussed in the chapter, to illustrate the adaptability and robustness of the element-free Galerkin method.

## REFERENCES


1. G.R. Liu, *Mesh Free Methods-Moving Beyond the Finite Element Method*, CRC Press (2003), ISBN: 9780849312380.
2. N. Moes, J. Dolbow, T. Belytschko, A finite element method for crack growth without remeshing, *International Journal for Numerical Methods in Engineering*, 46 (1999) 131–150.
3. T. Belytschko, T. Black, Elastic crack growth in finite elements with minimal remeshing, *International Journal for Numerical Methods in Engineering*, 45 (1999) 601–620.
4. W.K. Liu, S. Jun, Y.F. Zhang, Reproducing kernel particle methods, *International Journal for Numerical Methods in Engineering*, 20 (1995) 1081–1106.
5. S.N. Atluri, T. Zhu, A new meshless local Petrov-Galerkin (MLPG) approach in computational mechanics, *Computational Mechanics*, 22 (1998) 117–127.
6. T. Rabczuk, T. Belytschko, Cracking particles: A simplified meshfree method for arbitrary evolving cracks, *International Journal of Numerical Methods in Engineering*, 61 (2004) 2316–2343.
7. T. Belytschko, Y.Y. Lu, L. Gu, Element-free Galerkin methods, *International Journal for Numerical Methods in Engineering*, 37 (1994) 229–256.
8. R. Brighenti, Application of the element-free Galerkin meshless method to 3-D fracture mechanics problems, *Engineering Fracture Mechanics*, 72 (2005) 2808–2820.
9. M. Duflot, H. Dang, A truly Meshless Galerkin method based on a moving least squares quadrature, *Communications in Numerical Methods in Engineering*, 18 (2002) 441–449.
10. M. Fleming, Y.A. Chu, B. Moran, T. Belytschko, Enriched element free Galerkin methods for crack tip fields, *International Journal of Numerical Methods in Engineering*, 40 (1997) 1483–1504.
11. I.V. Singh, K. Sandeep, R. Prakash, The effect of weight function and scaling parameter on meshless EFG results in heat transfer problems, *International Journal of Heat & Technology*, 23 (2005) 13–20.
12. G. Yagawa, T. Furukawa, Recent developments of free mesh method, *International Journal for Numerical Methods in Engineering*, 47 (2000) 1419–1443.
13. L. Gavete, S. Falcon, J.C. Bellido, Dirichlet boundary conditions in element free Galerkin methods, in European Congress on Computational Methods in Applied Sciences and Engineering, Barcelona (2000) 1–14.
14. P. Nguyen, T. Rabczuk, S. Bordas, M. Duflot, Meshless methods: A review and computer implementation aspects, *Mathematics and Computers in Simulation*, 79 (2008) 763–813.
15. T. Belytschko, Y. Krongauz, M. Fleming, D. Organ, W.K. Liu, Smoothing and accelerated computations in element-free Galerkin method, *Journal of Computational and Applied Mathematics*, 74 (1996) 111–126.
16. J. Dolbow, T. Belytschko, An introduction to programming the meshless element free Galerkin method, *Archives in Computational Mechanics*, 5 (1998)207–241.
17. Y. Krongauz, T.Belytschko, Enforcement of essential boundary conditions in meshless approximations using finite elements. *Computer Methods in Applied Mechanics and Engineering*, 131 (1996) 133–145.

18. T.L. Anderson, *Fracture Mechanics: Fundamentals and Applications*, Taylor and Francis, CRC Press (2005).
19. T. Matsumoto, M. Tanaka, R. Obara, Computation of stress intensity factors of interface cracks based on interaction energy release rates and BEM sensitivity analysis, *Engineering Fracture Mechanics*, 65 (2000) 683–702.
20. X.Y. Liu, Q.Z. Xiao, B.L Karihaloo, XFEM for direct evaluation of mixed mode SIFs in homogeneous and bi-materials, *International Journal for Numerical Methods in Engineering*, 59 (2004) 1103–1118.
21. M. Duflot, The extended finite element method in thermo-elastic fracture mechanics, *International Journal for Numerical Methods in Engineering*, 74 (2008) 827–847.
22. N.N.V. Prasad, M.H. Aliabadi, D.P. Rooke, The dual boundary element method for thermo-elastic crack problems, *International Journal of Fracture*, 66 (1994) 255–272.

# 2 Evaporative Capillary Instability of Swirling Fluid Layer with Mass Transfer

*Mukesh Kumar Awasthi,*
*Rishi Asthana, and Ziya Uddin*

**CONTENTS**



## 2.1 INTRODUCTION

If a fluid cylinder collapses in an infinite fluid, such as atmospheric air, the instability at the interface is known as capillary instability. The interface becomes unstable due to the presence of capillary forces (a capillary force occurs due to surface tension at the free surface). This type of instability arises in various real-life phenomena, such as the breakup of a liquid jet, the situation of film boiling, and in various metallurgical and chemical processes.

The annular swirling flows have widespread applications, such as liquid rocket fuel injectors, swirlers, cylindrical cyclone separators, etc. The dynamics of swirling flows in a cylindrical configuration combine the complexity of swirl with pressure and surface tension, and therefore it is a challenging task to study the stability with a swirling cylindrical interface. The swirl activates a centrifugal force that induces an additional pressure gradient at the interface. Binnie and co-workers examined the waves propagated in a swirling liquid, along with surface tension [1]. The problem of co-axial injectors taking gas in the core with swirling was considered by Schumaker et al. [2], and the convective and absolute instabilities of annular liquid with swirling were studied by Fu and colleagues [3].

The outcome of heat/mass transfer on the interfacial stability of non-viscous fluids in a cylindrical shape has been measured by various authors [4–8]. Kim and colleagues used viscous irrotational theory to examine the heat/mass transfer effect on two fluid layers bounded in an annular region [9]. Awasthi and Agrawal added the nonlinear effects on the interfacial stability with heat/mass transfer at the vapor–liquid interface, bounded by co-axial cylinders [10]. The potential flow theory of viscous fluids has been extensively used by authors to study the interfacial instability of a cylindrical interface, allowing heat/mass transfer [11–14].

The consequence of swirl on the stability of an interface achieving heat/mass transfer was studied by Fu et al. [15, 16]. The swirl impact on temporal instability at the liquid/vapor interface was measured [15], and the Kelvin-Helmholtz instability at the cylindrical configuration was studied, including the effects of heat/mass transfer and free swirl [16]. The above investigations were restricted to non-viscous fluids, and the swirl effect was found to have a stabilizing impact.

Awasthi studied the Rayleigh-Taylor stability in the annular swirling layer with heat/mass transfer, and observed that swirling decays the perturbations [17]. Bo-qi and colleagues investigated the stability of confined swirling heat and mass transfer between annular liquids and coaxial jets, respectively, in the presence of velocity oscillations, and concluded that the velocity oscillations had a destabilizing effect on the overall flow [18, 19]. Stability analysis of heat and mass transfer flow in different annuli was studied by Awasthi et al. [20] and Qian et al. [21]. Awasthi [22] also investigated the effect of heat and mass transfer on the stability between a viscoelastic liquid and a viscous gas, and found that allowing heat transfer stabilized the system. To study the Rayleigh-Taylor instability of swirling annular fluid layers, a pressure correction model for the viscous potential flow was presented by Shivam and colleagues [23]. Awasthi examined the Kelvin-Helmholtz instability at the interface between a viscoelastic liquid and a viscous gas, and showed the impact of heat transfer in terms of the range of stable wave numbers and the maximum value of the growth rate parameter [24]. Recently, Awasthi and co-workers [25] investigated the temporal instability of the swirling annular layer and mass transfer in porous media, and reported a decrease in wave mode growth rate due to the presence of swirl.

In this chapter, the impact of swirl is examined on the instability of the interface between annular layers of viscous fluids. Heat/mass transfer has been achieved at the interface. The potential flow theory of viscous fluids, in which the fluid motion is considered to be irrotational, has been used to solve mathematical equations. The analysis given by Kim et al. was also based on irrotational fluid motion [9], but, unlike Kim and colleagues, we do not set the rotation of the outer cylinder to zero. Therefore, the centrifugal force enters into the analysis. Furthermore, the interfacial stability of the vapor–liquid interface in an annular region with heat and mass transfer was examined by Fu and co-workers [15], but their analysis did not include viscous effects. In this chapter, we consider that the stability of the vapor–liquid interface, bounded by two concentric cylinders and viscous effects, is not zero. Therefore, the viscous normal stresses enter into the study; however, the effect of gravity and flow in the axial direction is not taken into account in the analysis. To observe the amplitude of the perturbations, the normal mode procedure is used and a critical wave number, depending on the physical parameters, is attained. The effects of numerous dimensionless variables on the stability are explored in this investigation.

## 2.2 MATHEMATICAL DESCRIPTION

The geometrical configuration displayed in Figure 2.1 is a swirling annular viscous fluid layer enclosed by two cylinders of radii $a$, $b$ ($a < b$), with vapor lying in the inner phase. The sub suffix ($i$) is adopted to represent the inner phase, while ($o$) denotes the outer one. The inner and outer fluid phases have viscosities $\mu_i$ and $\mu_o$, densities $\rho_i$ and $\rho_o$, and temperatures $T_i$ and $T_o$, respectively, while $\sigma$ denotes the surface tension of the fluid. The outer cylinder is swirling anticlockwise with velocity $r\Omega$, where $\Omega$ represents the angular velocity of the cylinder. The governing equations of the fluid phases are given as follows:

$$\left.\begin{aligned}
&\frac{1}{r}\frac{\partial(ru_i)}{\partial r}+\frac{1}{r}\frac{\partial v_i}{\partial\theta}+\frac{\partial w_i}{\partial z}=0\\
&\frac{\partial u_i}{\partial t}+u_i\frac{\partial u_i}{\partial r}+\frac{v_i}{r}\frac{\partial u_i}{\partial\theta}+w_i\frac{\partial u_i}{\partial z}-\frac{v_i^2}{r}=-\frac{1}{\rho_i}\frac{\partial p_i}{\partial r}+\nu_i\left(\nabla^2u_i-\frac{u_i}{r^2}-\frac{2}{r^2}\frac{\partial v_i}{\partial\theta}\right)\\
&\frac{\partial v_i}{\partial t}+u_i\frac{\partial v_i}{\partial r}+\frac{v_i}{r}\frac{\partial v_i}{\partial\theta}+w_i\frac{\partial v_i}{\partial z}+\frac{u_iv_i}{r}=-\frac{1}{\rho_i}\frac{1}{r}\frac{\partial p_i}{\partial\theta}+\nu_i\left(\nabla^2v_i+\frac{2}{r^2}\frac{\partial u_i}{\partial\theta}-\frac{v_i}{r^2}\right)\\
&\frac{\partial w_i}{\partial t}+u_i\frac{\partial w_i}{\partial r}+\frac{v_i}{r}\frac{\partial w_i}{\partial\theta}+w_i\frac{\partial w_i}{\partial z}=-\frac{1}{\rho_i}\frac{\partial p_i}{\partial z}+\nu_i\left(\nabla^2w_i\right)
\end{aligned}\right\}\quad(2.1)$$

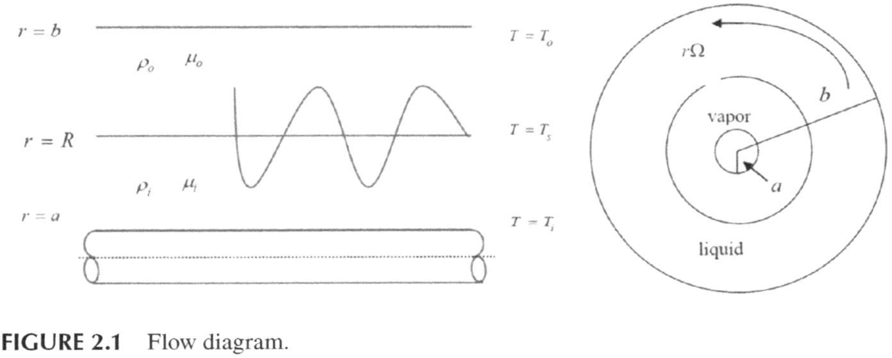


**FIGURE 2.1** Flow diagram.

$$\left.\begin{aligned}
&\frac{1}{r}\frac{\partial(ru_o)}{\partial r}+\frac{1}{r}\frac{\partial v_o}{\partial \theta}+\frac{\partial w_o}{\partial z}=0\\
&\frac{\partial u_o}{\partial t}+u_o\frac{\partial u_o}{\partial r}+\frac{v_o}{r}\frac{\partial u_o}{\partial \theta}+w_o\frac{\partial u_o}{\partial z}-\frac{v_o^2}{r}=-\frac{1}{\rho_o}\frac{\partial p_o}{\partial r}+\nu_o\left(\nabla^2 u_o-\frac{u_o}{r^2}-\frac{2}{r^2}\frac{\partial v_o}{\partial \theta}\right)\\
&\frac{\partial v_o}{\partial t}+u_o\frac{\partial v_o}{\partial r}+\frac{v_o}{r}\frac{\partial v_o}{\partial \theta}+w_o\frac{\partial v_o}{\partial z}+\frac{u_o v_o}{r}=-\frac{1}{\rho_o}\frac{1}{r}\frac{\partial p_o}{\partial \theta}+\nu_o\left(\nabla^2 v_o+\frac{2}{r^2}\frac{\partial u_o}{\partial \theta}-\frac{v_o}{r^2}\right)\\
&\frac{\partial w_o}{\partial t}+u_o\frac{\partial w_o}{\partial r}+\frac{v_o}{r}\frac{\partial w_o}{\partial \theta}+w_o\frac{\partial w_o}{\partial z}=-\frac{1}{\rho_o}\frac{\partial p_o}{\partial z}+\nu_o\left(\nabla^2 w_o\right)
\end{aligned}\right\} \tag{2.2}$$

where $(u_i, v_i, w_i)$ and $(u_o, v_o, w_o)$ denote the velocity in the inner and outer phases, respectively, and $\nabla^2=\frac{\partial^2}{\partial r^2}+\frac{1}{r^2}\frac{\partial}{\partial \theta^2}+\frac{1}{r}\frac{\partial}{\partial r}+\frac{\partial^2}{\partial z^2}$.

At the interface, the mass is transporting one phase to the other phase. In mathematical form, we write

$$\rho_i\left(\frac{df}{dt}+\boldsymbol{v}_i\cdot\nabla f\right)=\rho_o\left(\frac{df}{dt}+\boldsymbol{v}_o\cdot\nabla f\right) \tag{2.3}$$

If the fluids are thermally conducting with a thermal diffusivity $D_T$, the equation of diffusion can be written as

$$\frac{\partial T}{\partial t}+\boldsymbol{v}\cdot\nabla T=D_T\nabla^2 T \tag{2.4}$$

We assume that the inner fluid has a temperature $T_i$ and the outer fluid has a temperature $T_o$. Also, the fluids are in thermodynamic equilibrium in the basic situation and the saturation temperature is $T_s$.

The heat transfer treatment is analyzed, accepting the model of Awasthi [17]. The total heat flux is equated to the latent heat and, therefore, the diffusion equation (2.4) takes the form as

$$L\rho_i\left(\frac{\partial f}{\partial t}+\boldsymbol{v}_i\cdot\nabla f\right)=H(r)=\frac{\kappa_o(T_s-T_o)}{r[\ln b-\ln r]}-\frac{\kappa_i(T_i-T_s)}{r[\ln r-\ln a]} \tag{2.5}$$

where $L$ and $H(r)$ represent the latent heat and total heat flux, respectively. Here, $\kappa_i$ and $\kappa_o$ denote the thermal conductivities of the inner and outer phases, respectively.

The stress equilibrium at the interface is written as

$$\left(p_o+\frac{1}{2}\rho_o\Omega^2 r^2-2\mu_o\hat{\boldsymbol{n}}\cdot\nabla\otimes\boldsymbol{v}_o\cdot\hat{\boldsymbol{n}}\right)-\left(p_i-2\mu_i\hat{\boldsymbol{n}}\cdot\nabla\otimes\boldsymbol{v}_i\cdot\hat{\boldsymbol{n}}\right)=\sigma\nabla\cdot\hat{\boldsymbol{n}} \tag{2.6}$$

where $\hat{\boldsymbol{n}}$ is the outward unit normal.

### 2.2.1 Basic State

The basic state is given by $r = R$ and, therefore, the velocities in the inner phase and the outer phase are given by (0, 0, 0) and (0, $r\Omega$, 0), respectively, and the inner and outer phase pressures are denoted by $P_i$ and $P_o$, respectively. Using these values, Eqs. (2.1) and (2.2) take the form of:

$$\frac{\partial P_i}{\partial r} = \frac{\partial P_i}{\partial \theta} = \frac{\partial P_i}{\partial z} = 0 \tag{2.7}$$

$$\frac{\partial P_o}{\partial r} = \rho_o \Omega^2 R; \quad \frac{\partial P_o}{\partial \theta} = \frac{\partial P_o}{\partial z} = 0 \tag{2.8}$$

There will no mass transfer in the basic state.

The total heat flux at every phase is the same in an equilibrium state and hence,

$$\frac{\kappa_o(T_s - T_o)}{R[\ln b - \ln R]} = \frac{\kappa_i(T_i - T_s)}{R[\ln R - \ln a]} = \frac{(T_i - T_o)}{[\ln(R/a)/\kappa_i + \ln(b/R)/\kappa_o]} = Q(\text{say}) \tag{2.9}$$

The interfacial stress balance reduces to

$$\left( P_o + \frac{1}{2}\rho_o \Omega^2 R^2 \right) = P_i \tag{2.10}$$

### 2.2.2 Perturbed State

To study the stability of the interface, a small distortion is applied to the interface and it takes the form $r = R + \xi(\theta, z, t)$. The fluid flow field is also distorted with the disturbed pressure and flow velocity, i.e., $\left[u_i, v_i, w_i, p_i, u_o, v_o, w_o, p_o\right] = \left[\bar{u}_i, \bar{v}_i, \bar{w}_i, P_i + \bar{p}_i, \bar{u}_o, r\Omega + \bar{v}_o, \bar{w}_o, P_o + \bar{p}_o\right]$. Using these values in the governing equations (2.1) and (2.2), retaining only the linear terms and removing the bar sign, we have:

$$\left.\begin{aligned}
&\frac{1}{r}\frac{\partial(ru_i)}{\partial r} + \frac{1}{r}\frac{\partial v_i}{\partial \theta} + \frac{\partial w_i}{\partial z} = 0 \\
&\frac{\partial u_i}{\partial t} = -\frac{1}{\rho_i}\frac{\partial p_i}{\partial r} + \nu_i\left( \nabla^2 u_i - \frac{u_i}{r^2} - \frac{2}{r^2}\frac{\partial v_i}{\partial \theta} \right) \\
&\frac{\partial v_i}{\partial t} = -\frac{1}{\rho_i}\frac{1}{r}\frac{\partial p_i}{\partial \theta} + \nu_i\left( \nabla^2 v_i + \frac{2}{r^2}\frac{\partial u_i}{\partial \theta} - \frac{v_i}{r^2} \right) \\
&\frac{\partial w_i}{\partial t} = -\frac{1}{\rho_i}\frac{\partial p_i}{\partial z} + \nu_i\left(\nabla^2 w_i\right)
\end{aligned}\right\} \tag{2.11}$$

$$\left.\begin{aligned}
&\frac{1}{r}\frac{\partial(ru_o)}{\partial r}+\frac{1}{r}\frac{\partial v_o}{\partial\theta}+\frac{\partial w_o}{\partial z}=0\\
&\left(\frac{\partial u_o}{\partial t}+\Omega\frac{\partial u_o}{\partial\theta}-2\Omega v_o\right)=-\frac{1}{\rho_o}\frac{\partial p_o}{\partial r}+v_o\left(\nabla^2u_o-\frac{u_o}{r^2}-\frac{2}{r^2}\frac{\partial v_o}{\partial\theta}\right)\\
&\left(\frac{\partial v_o}{\partial t}+\Omega\frac{\partial v_o}{\partial\theta}+2\Omega u_o\right)=-\frac{1}{\rho_o}\frac{1}{r}\frac{\partial p_o}{\partial\theta}+v_o\left(\nabla^2v_o+\frac{2}{r^2}\frac{\partial u_o}{\partial\theta}-\frac{v_o}{r^2}\right)\\
&\left(\frac{\partial w_o}{\partial t}+\Omega\frac{\partial w_o}{\partial\theta}\right)=-\frac{1}{\rho_o}\frac{\partial p_o}{\partial z}+v_o\left(\nabla^2w_o\right)
\end{aligned}\right\}\tag{2.12}$$

The equation of the mass transport at the perturbed interface becomes

$$\rho_i\left(u_i-\frac{\partial\xi}{\partial t}\right)=\rho_o\left(u_o-\frac{\partial\xi}{\partial t}\right)\tag{2.13}$$

The linear heat transport equation at the perturbed interface is described by:

$$L\rho_i\left(u_i-\frac{\partial\xi}{\partial t}\right)=H(R+\xi)\tag{2.14}$$

In the linear case $H(R+\xi)=H(R)+\xi H'(R)$ and $H(R)=0$. Hence, $H(R+\xi)=\xi H'(R)$. Now, Eq. (2.14) takes the form:

$$\rho_i\left(u_i-\frac{\partial\xi}{\partial t}\right)=\alpha\xi\tag{2.15}$$

where $\alpha=\dfrac{H'(R)}{L}$

Also, the radial velocities $u_i$, $u_o$ vanish at the rigid boundaries $r=a, b$, respectively.

The normal mode analysis is used to examine the stability. The interface deviation $\xi(\theta,z,t)$ may be given in the form of sinusoidal waves as

$$\xi=\hat{\xi}\,e^{i(kz+m\theta)}\,e^{-i\omega t}+c.c.\tag{2.16}$$

The perturbed fluid motion is considered to be irrotational and so $\boldsymbol{v_i}=(u_i,v_i,w_i)=\nabla\varphi_i$ and $\boldsymbol{v_o}=(u_o,v_o,w_o)=\nabla\varphi_o$. Assuming $\varphi(r,\theta,z,t)=H(r)e^{i(kz+m\theta)}\,e^{-i\omega t}+c.c.$ and using the conditions from Eqs. (2.13) and (2.15), we have

$$\varphi_i=\frac{1}{k}\left(\frac{\alpha}{\rho_i}-i\omega\right)A_i(kr)\xi\,e^{i(kz+m\theta)}\,e^{-i\omega t}+c.c.\tag{2.17}$$

$$\varphi_o=\frac{1}{k}\left(\frac{\alpha}{\rho_o}-i\omega\right)A_o(kr)\xi\,e^{i(kz+m\theta)}\,e^{-i\omega t}+c.c.\tag{2.18}$$

Here

$$A_i(kr) = \frac{I'_m(ka)K_m(kr) - I_m(kr)K'_m(ka)}{I'_m(ka)K'_m(kR) - I'_m(kR)K'_m(ka)}, \quad A_o(kr) = \frac{I'_m(kb)K_m(kr) - I_m(kr)K'_m(kb)}{I'_m(kb)K'_m(kR) - I'_m(kR)K'_m(kb)}$$

where $K_m(kr)$ and $I_m(kr)$ have the usual meaning.

In this analysis, the viscosity is assumed to be non-zero. The interfacial dynamic equation, which includes viscosity, is written as:

$$p_o + \rho_o\Omega^2 R\xi - p_i - 2\mu_o\frac{\partial^2\varphi_o}{\partial r^2} + 2\mu_i\frac{\partial^2\varphi_i}{\partial r^2} = \sigma\left(\frac{\partial^2\xi}{\partial z^2} + \frac{1}{R^2}\frac{\partial^2\xi}{\partial\theta^2} + \frac{\xi}{R^2}\right) \tag{2.19}$$

The well-known Bernoulli's equation has been used to calculate the pressure at the free surface. Eq. (2.19) can be re-written as

$$\rho_o\left(\frac{\partial\varphi_o}{\partial t}\right) - \rho_o\Omega^2 R\xi + 2\mu_o\frac{\partial^2\varphi_o}{\partial r^2} - \rho_i\left(\frac{\partial\varphi_i}{\partial t}\right) - 2\mu_i\frac{\partial^2\varphi_i}{\partial r^2} = -\sigma\left(\frac{\partial^2\xi}{\partial z^2} + \frac{1}{R^2}\frac{\partial^2\xi}{\partial\theta^2} + \frac{\xi}{R^2}\right) \tag{2.20}$$

The expressions $\varphi_i$, $\varphi_o$ are used in Eq. (2.20) and we obtain a quadratic equation in $\omega$ as:

$$b_1\omega^2 + ib_2\omega - b_3 = 0 \tag{2.21}$$

where $b_1 = \rho_i A_i(kR) - \rho_o A_o(kR)(kR)$

$$b_2 = \alpha\left(A_i(kR) - A_o(kR)\right) + 2k^2\left(\mu_i B_i(kR) - \mu_o B_o(kR)\right)$$

$$b_3 = \frac{\sigma k}{R^2}\left(k^2R^2 + m^2 - 1\right) + 2k^2\alpha\left(\frac{\mu_i}{\rho_i}B_i(kR) - \frac{\mu_o}{\rho_o}B_o(kR)\right) + \rho_o\Omega^2 Rk$$

$$B_j(kR) = \left(1 + \frac{m^2}{k^2R^2}\right)A_j(kR) - \frac{1}{kR}, \quad (j = i, o)$$

The parameter $\omega$ is assumed to be complex, i.e., $\omega = \omega_R + i_{\omega I}$. Putting in the value of $\omega$ and writing the real and imaginary parts, we have:

$$2b_1\omega_R\omega_I + b_2\omega_R = 0 \Rightarrow \omega_R = 0 \tag{2.22}$$

$$\text{and } b_1(\omega_R^2 - \omega_I^2) - b_2\omega_I + b_3 = 0 \tag{2.23}$$

The value of $\omega_R$ obtained in Eq. (2.22) is used in Eq. (2.23) and, hence, Eq. (2.23) reduces, in terms of $\omega_I$, as:

$$\begin{aligned}&\left[\rho_i A_i(kR)-\rho_o A_o(kR)\right]\omega_I^2\\&+\left[\alpha\left(A_i(kR)-A_o(kR)\right)+2k^2\left(\mu_i B_i(kR)-\mu_o B_o(kR)\right)\right]\omega_I\\&+\left[\frac{\sigma k}{R^2}\left(k^2R^2+m^2-1\right)+2k^2\alpha\left(\frac{\mu_i}{\rho_i}B_i(kR)-\frac{\mu_o}{\rho_o}B_o(kR)\right)+\rho_o\Omega^2 Rk\right]=0\end{aligned} \tag{2.24}$$

If we put the fluid viscosities $\mu_i$ and $\mu_o$ equal to zero and both cylinders are fixed ($\Omega = 0$), Eq. (2.24) reduces to the same equation as achieved by Lee [6].

For the stable arrangement, $\omega_I$ should be negative and, to calculate negative roots of Eq. (2.24), the Routh-Hurwitz criteria are applied and conditions for negative $\omega_{I,}$ i.e., for stability, are given as $b_1 > 0$, $b_2 > 0$, $b_3 > 0$.

The parameters $\rho_i$, $\rho_o$, $\mu_i$, $\mu_o$, and $\alpha$ are all positive and, checking the values of Bessel's functions, we have $b_1 > 0$, $b_2 > 0$ trivially and therefore, the condition for stable arrangement has been reduced to $b_3 > 0$. The condition for the marginal stable state is given by $b_3 = 0$.

$$\begin{aligned}\text{i.e. } &\frac{\sigma}{R^2}\left(k^2R^2+m^2-1\right)+2k\alpha\left(\frac{\mu_i}{\rho_i}B_i(kR)-\frac{\mu_o}{\rho_o}B_o(kR)\right)\\&+\rho_o\Omega^2 R=0\end{aligned} \tag{2.25}$$

Here, $m \geq 0$. For $m \geq 1$, Eq. (2.25) does not change sign, i.e., for asymmetric disturbances, there is no positive wave number $k$ and, hence, the liquid/vapor interface is always stable for asymmetric disturbances.

## 2.3 DIMENSIONLESS FORM OF THE DISPERSION RELATIONSHIP

Let us assume that $\tau = \left[\rho_o H^3/\sigma\right]^{1/2}$, and $H = b - a$ represent characteristic time and characteristic length, respectively. The other variables in dimensionless form are:

$$\hat{k} = kH \qquad \hat{\alpha} = \frac{\alpha\tau}{\rho_o} \qquad \hat{h}_1 = \frac{h_1}{H} = \varphi \qquad \hat{h}_2 = \frac{h_2}{H} = 1-\hat{h}_1 \qquad \rho = \frac{\rho_i}{\rho_o}$$

$$\mu = \frac{\mu_i}{\mu_o}, \qquad \kappa = \frac{\mu}{\rho}, \qquad \text{Oh} = \frac{\sqrt{\rho_o\sigma H}}{\mu_o}, \qquad \Lambda = \frac{2\hat{\alpha}}{\text{Oh}}, \qquad Ce = \frac{\rho_o\Omega^2 H^3}{\sigma}$$

Here, $Ce$ is the centrifuge number. The dimensionless form of Eqs. (2.24) and (2.25) are:

$$\left[\rho A_i(\hat{k}\hat{R}) - A_o(\hat{k}\hat{R})\right]\hat{\omega}_I^2$$
$$+\left[\hat{\alpha}\left(A_i(\hat{k}\hat{R}) - A_o(\hat{k}\hat{R})\right) + \frac{2\hat{k}^2}{Oh}\left(\mu B_i(\hat{k}\hat{R}) - B_o(\hat{k}\hat{R})\right)\right]\hat{\omega}_I \tag{2.26}$$
$$+\left[\frac{\hat{k}}{\hat{R}^2}\left(\hat{k}^2\hat{R}^2 + m^2 - 1\right) + \Lambda\hat{k}^2\left(\kappa B_i(\hat{k}\hat{R}) - B_o(\hat{k}\hat{R})\right) + \hat{k}\hat{R}Ce\right] = 0$$

$$\frac{1}{\hat{R}^2}\left(\hat{k}^2\hat{R}^2 + m^2 - 1\right) + \Lambda\hat{k}\left(\kappa B_i(\hat{k}\hat{R}) - B_o(\hat{k}\hat{R})\right) + \hat{R}Ce = 0 \tag{2.27}$$

If we do not allow heat/mass transport at the interface, i.e., $\Lambda = 0$, the viscosity will not affect the criterion for stability. By setting the centrifuge number, i.e., $Ce \to 0$, Eq.(2.26) shortens to

$$\left[\rho A_i(\hat{k}\hat{R}) - A_o(\hat{k}\hat{R})\right]\hat{\omega}_I^2$$
$$+\left[\hat{\alpha}\left(A_i(\hat{k}\hat{R}) - A_o(\hat{k}\hat{R})\right) + \frac{2\hat{k}^2}{\mathrm{Oh}}\left(\mu B_i(\hat{k}\hat{R}) - B_o(\hat{k}\hat{R})\right)\right]\hat{\omega}_I \tag{2.28}$$
$$+\left[\frac{\hat{k}}{\hat{R}^2}\left(\hat{k}^2\hat{R}^2 + m^2 - 1\right) + \Lambda\hat{k}^2\left(\kappa B_i(\hat{k}\hat{R}) - B_o(\hat{k}\hat{R})\right)\right] = 0$$

The same expression is achieved by Kim et al. [9] for axisymmetric distortion, i.e., $m = 0$, and Awasthi et al. [10] for their linear theory.

For small wavelength disturbance waves, i.e., $k \to \infty$, Eq. (2.27) is given as:

$$\hat{k}^2 + \Lambda\hat{k}\left(\kappa + 1\right) + \hat{R}Ce = 0 \tag{2.29}$$

In this case, $\hat{k}_c$ is always negative because both $\Lambda$ and $\kappa$ are positive.

In the case of large wavelength disturbance waves ($k \to 0$), Eq. (2.27) is given as:

$$\left(m^2 + \hat{k}^2\hat{R}^2 - 1\right) + \Lambda\hat{R}^2\left(\kappa\left(\frac{2\hat{R}}{\hat{R}^2 - a^2} - \frac{1}{\hat{R}}\right) - \left(\frac{2\hat{R}}{b^2 - \hat{R}^2} - \frac{1}{\hat{R}}\right)\right) + \hat{R}^3 Ce = 0 \tag{2.30}$$

In this case, there will be no critical wave number because

$$\Lambda\left(\kappa\left(\frac{2\hat{R}}{\hat{R}^2 - a^2} - \frac{1}{\hat{R}}\right) - \left(\frac{2\hat{R}}{b^2 - \hat{R}^2} - \frac{1}{\hat{R}}\right)\right) + \hat{R}Ce > 1 \tag{2.31}$$

## 2.4 NUMERICAL RESULTS AND DISCUSSIONS

This section is committed to indicating the outcomes of various physical variables of the considered problem, on the stability principle. The numerical computation has been made for Eqs. (2.26) and (2.27). For the asymmetric perturbation, we found that the system is always stable, and, therefore, the plots are made for axisymmetric mode, i.e., $m = 0$. Heat always transfers from high temperature to low temperature. In this case, the vapor is hot as compared with liquid, and the vapor is also miscible in the liquid. At the interface, where the vapor mixes with the liquid, heat/mass transfer comes into the picture. Hence, the combined heat/mass transfer effect has been studied with swirling. The case of film boiling has been studied in this analysis and, therefore, water lies in the outer phase, whereas vapor lies in the inner phase. The stable and unstable regions are referred to in the figures. The physical variable values used in the numerical computation are given as follows:

$$\rho_i = 0.001 \text{ gm/cm}^3,\ \mu_i = 0.00001 \text{ poise},\ a = 0.5 \text{ cm},\ b = 1.0 \text{ cm}$$

$$\rho_o = 1.0 \text{ gm/cm}^3,\ \mu_o = 0.01 \text{poise},\ \sigma = 72.3 \text{ dyne/cm}$$

The coupled heat/mass transport impact on the liquid/vapor interface in an annular configuration was analyzed by Kim et al. [9]. They considered the motion to be irrotational and the fluids are viscous. The present analysis includes the swirling effect at the liquid/vapor interface, allowing heat/mass transport. To study the swirl behavior, the range of stability obtained in the present analysis is compared with the range obtained by Kim and colleagues [9] in Figure 2.2. As we have considered that the liquid phase is swirling, a greater amount of centrifugal force will be encountered in this phase and this force will affect it at a speed away from the center of curvature, whereas vapor will tend toward the center of curvature. The centrifugal force neutralizes the perturbation and balances out the interface; therefore, the addition of swirl enlarges the range of stability, as observed from Figure 2.2.

To explain the influence of centrifuge numbers on the stability range, Figure 2.3 is generated by solving Eq. (2.27). As shown in this plot, as the centrifuge number *Ce* increases, the range of stability also increases. This shows a stabilizing influence of this parameter. The centrifuge number is the ratio of rotational force to surface tension. As the centrifuge number increases, the swirl velocity increases, or the surface tension decreases. The centrifugal force caused by the swirling movement neutralizes the development of the perturbation growth. The stability improvement mechanism for the interface increases the aerodynamic effects, which oppose the development of the magnitude of the perturbation caused by the surface tension and, therefore, the magnitude of the perturbation on the liquid/vapor interface decreases.

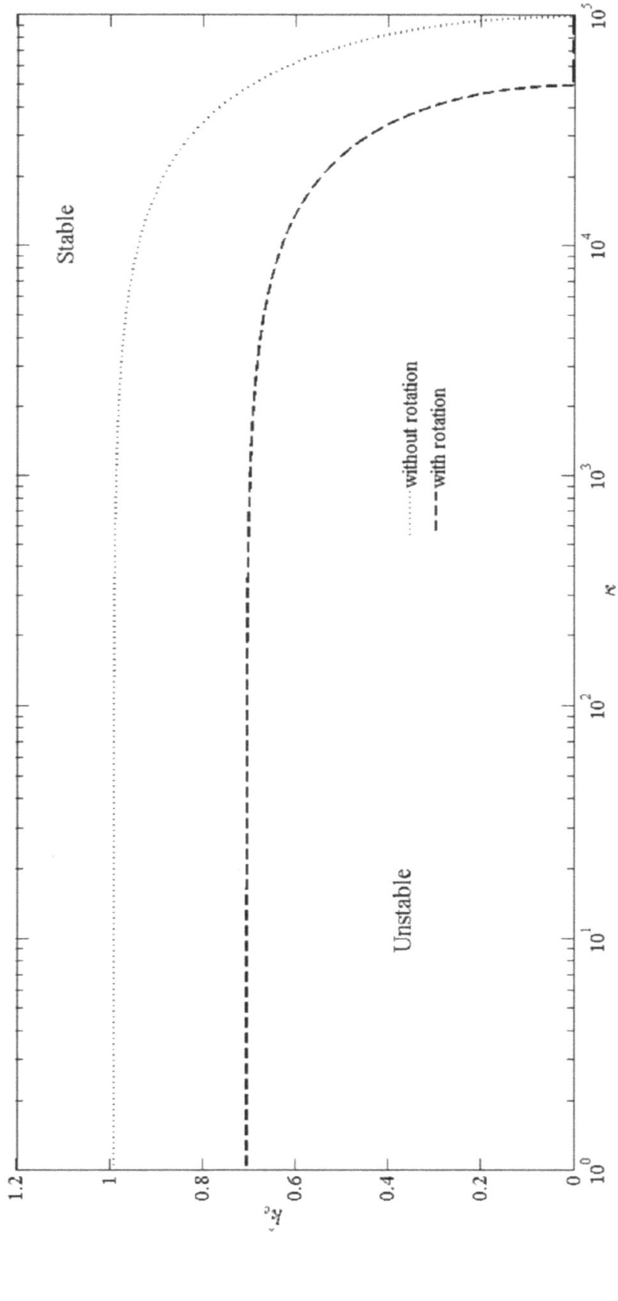


**FIGURE 2.2** Results comparing Eqs. (2.18) and (2.19) ($Ce = 0.0$ and 0.5, respectively).

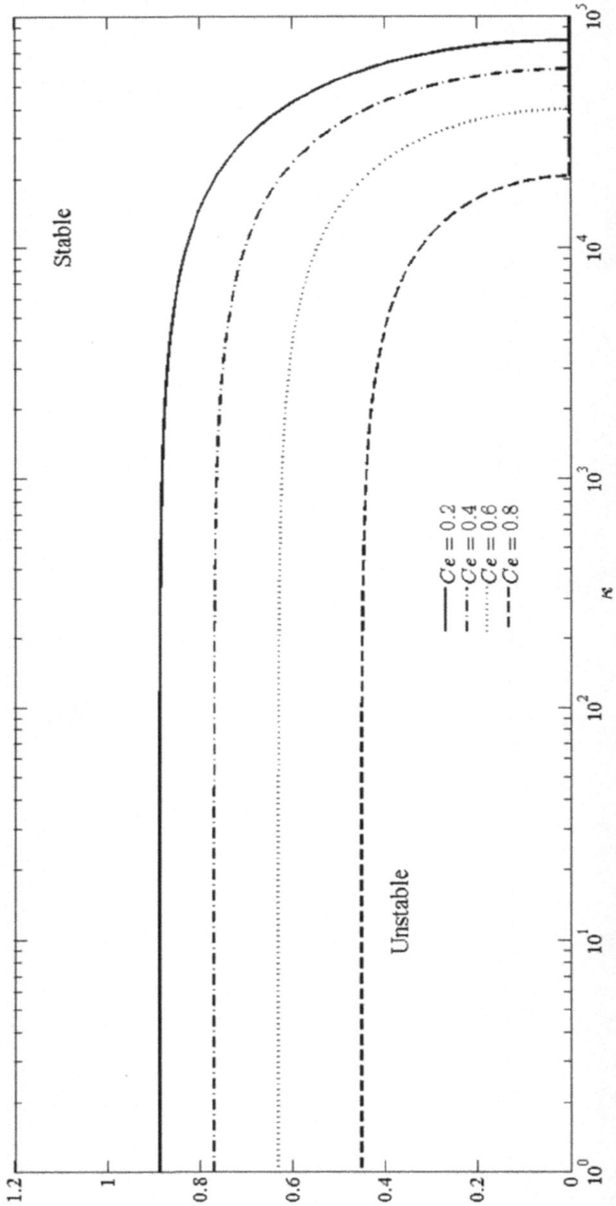


**FIGURE 2.3** Effect of centrifuge number $Ce$ ($\varphi = 0.01$, $\Lambda = 10^{-7}$).

Figure 2.4 shows that the outcome of heat/mass transport is not affected by the addition of swirl at the interface between two fluids. In this figure, the marginal stability condition, Eq. (2.27), has been plotted for different values of the amount of alternative heat-transfer constant $\Lambda$. The stable area increases considerably when the number of $\Lambda$ rises. The coefficient $\Lambda$ is the fraction between heat flux and the Ohnesorge number. This observation affirms that mass, as well as heat transfer, has a stabilizing effect, whereas the Ohnesorge number has a destabilizing effect. Similar results for the liquid/vapor interface in the cylindrical configuration were achieved by Kim et al. [9] and Awasthi [10–12]. We can clarify the effect of mass exchange on the interface, relative to evaporation and condensation. As the peaks are closest to the hotter vapor region, they are hotter than the troughs, since the troughs are closer to the cooler region. Evaporation will occur at the peaks and there will be condensation at the troughs. The fluid is stretching out toward the hotter region and, as a consequence, more evaporation will happen. This additional evaporation neutralizes the perturbation and the interface becomes more stable.

The oscillation of the critical amount of wave number $\hat{k}_c$, concerning the alternative heat transfer constant $\Lambda$ for a range of values of the kinematic viscosity ratio $\kappa$, is shown in Figure 2.5. It is noted that, as $\kappa$ increases, the stability range also increases. As $\kappa = \mu_1\rho_2/\mu_2\rho_1$, we may infer that the vapor viscosity demonstrates a stabilizing role, whereas liquid viscosity has a destabilizing impact. Liquid density opposes the growth of perturbations but vapor density supports the development of disturbance waves at the free surface.

The growth rate variation of perturbations $\omega_I$ concerning centrifuge number $Ce$ has been shown in Figure 2.6. The centrifuge number varies from 0.0 to 0.3, while other variables are constant. As shown in the figure, as $Ce$ increases, the growth rate decreases. The swirl generates centrifugal force at the interface and this force prevents the development of perturbations.

## 2.5 CONCLUSIONS

In the current study, the linear capillary stability of rotating annular layers has been viewed with respect to when heat/mass transport is taking place at the interface. We compute an analytical relationship, which is a polynomial of degree 2, with the growth rate parameter. The plots of the marginal state and the growth rate parameter are presented, depicting the effect of various physical parameters. With an increase in centrifuge number, the growth of perturbations decreases, and the stable range increases, concluding that, if annular layers are swirling, the interface moves toward stability. Moreover, the mass, as well as the heat transfer, also makes an impact on the interface by neutralizing the perturbation. Vapor viscosity and liquid density resist the growth of perturbation, whereas vapor density and liquid viscosity both have the opposite effect.

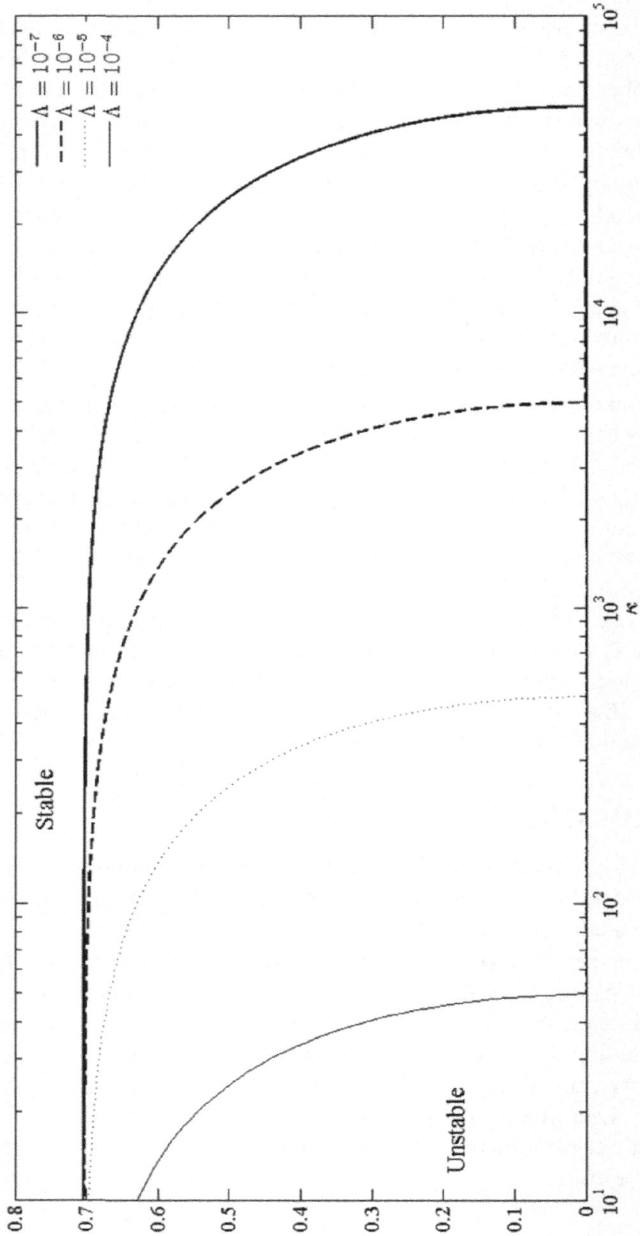


**FIGURE 2.4** Effect of alternative heat transfer constant $\Lambda$ ($\varphi = 0.01$, $Ce = 0.5$).

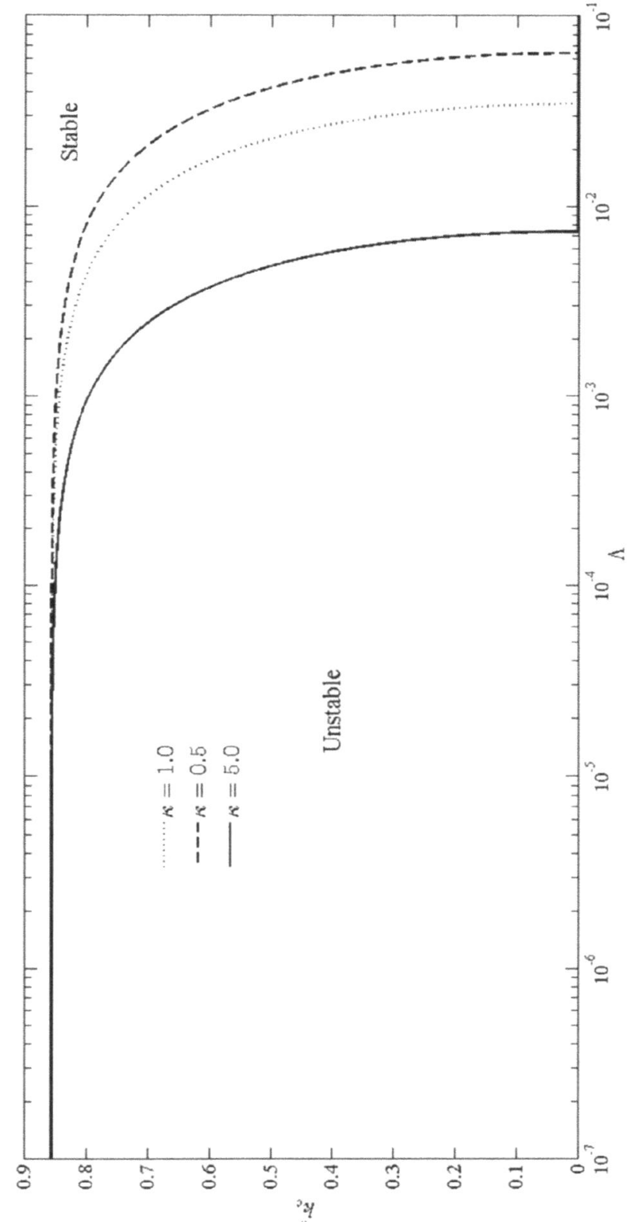


**FIGURE 2.5** Effect of kinematic viscosity ratio $\kappa$ ($\varphi = 0.05$, $Ce = 0.2$).

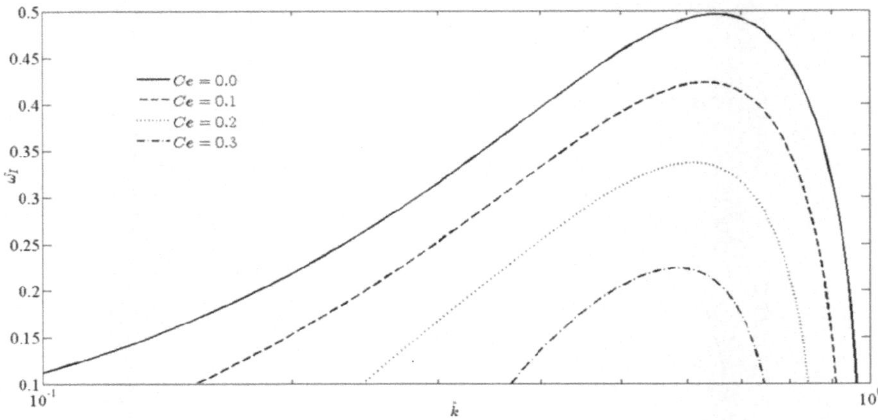


**FIGURE 2.6** Effect of swirl on growth rate ($\Lambda = 10^{-7}$, $\varphi = 0.05$).

## ACKNOWLEDGMENT

The author M.K.A. is thankful to University Grant Commission-Basic Scientific Research (UGC-BSR) (New-Delhi, India) (project No.F.30-442/2018 (BSR)) for a start-up grant and research funding.

## REFERENCES

1. Binnie, A.M. Theory of waves traveling on the core in a swirling liquid. *Proceedings of the Royal Society of London A*, 205 (1951), 530–540.
2. Schumaker, S.A., Danczyk, S., and Lightfoot, M. Effect of swirl on gas-centered swirl-coaxial injectors. In 47th AIAA/ASME/SAE/ASEE Joint Propulsion Conference & Exhibit, AIAA Paper 5621, San Diego, CA, 2011.
3. Fu, Q.F., Yang, L.J., Tong, M.X., and Wang, C. Absolute and convective instability of a confined swirling annular liquid layer. *Atomization and Sprays*, 24 (2014), 555–573.
4. Nayak, A.R., and Chakraborty, B.B. Kelvin-Helmholtz stability with mass and heat transfer. *Physics of Fluids*, 27 (1984) 1937.
5. Lee, D.S. Nonlinear instability of cylindrical flow with mass and heat transfer. *Physica Scripta*, 67 (2002) 420–426.
6. Lee, D.S. Nonlinear Rayleigh instability of cylindrical flow with mass and heat transfer. *Journal Physics A: Mathematical and General*, 36 (2003) 573–580.
7. Lee, D.S. Nonlinear Kelvin-Helmholtz instability of cylindrical flow with mass and heat transfer. *Physica Scripta*, 76 (2007) 97–103.
8. Seadawy, A.R., and El-Rashidy, K. Nonlinear Rayleigh–Taylor instability of the cylindrical fluid flow with mass and heat transfer. *Pramana*, 87 (2016) 20.
9. Kim, H. J., Kwon, S. J., Padrino, J. C. and Funada, T., "Viscous potential flow analysis of capillary instability with heat and mass transfer", *Journal of Physics A: Mathematical and Theoretical*. 41 (2008) 1–10.
10. Awasthi, M.K., and Agrawal, G.S. Nonlinear analysis of capillary instability with heat and mass transfer. *Communications in Nonlinear Science and Numerical Simulation*, 17 (2012) 2463–2475.

11. Awasthi, M.K. Nonlinear analysis of Rayleigh–Taylor instability of cylindrical flow with heat and mass transfer. *ASME Journal of Fluids Engineering*, 135 (2013) 061205 (7 pages).
12. Awasthi, M.K., and Asthana, R. Viscous potential flow analysis of capillary instability with and mass transfer through porous media. *International Communications in Heat and Mass Transfer*, 41 (2013) 7–11.
13. Asthana, R., Awasthi, M.K. and Agrawal, G.S. Viscous potential flow analysis of Kelvin–Helmholtz instability of cylindrical flow with heat and mass transfer. *Heat Transfer: Asian Research*, 43 (2014) 489–503.
14. Awasthi, M.K., Asthana, R., and Uddin, Z. Nonlinear study of Kelvin–Helmholtz instability of cylindrical flow with mass and heat transfer. *International Communications in Heat and Mass Transfer*, 71(2016), 216–224.
15. Fu, Q.F., Jia, B.Q., and Yang, L.J. Stability of a Confined SwirlingAnnular Liquid Layer with Heat and Mass Transfer. *InternationalJournal of Heat and Mass Transfer*, 104 (2017), 644–649.
16. Fu, Q.F., Deng, X.D., Jia, B.Q., and Yang, L.J. Temporal instability of a confined liquid film with heat and mass transfer. *AIAA Journal*, 56(7) (2018), 2615–2622. DOI: https://doi.org/10.2514/1.J056834.
17. Awasthi, M.K. Rayleigh-Taylor instability of swirling annular layer with mass transfer. *ASME Journal of Fluids Engineering*, 141 (2019) 071202 (5 pages).
18. Bo-qi Jia, Li-jun Yang, Luo Xie, Qing-fei Fu, and Xiao Cui. Linear stability of confined swirling annular liquid layers in the presence of gas velocity oscillations with heat and mass transfer. *International Journal of Heat and Mass Transfer*, 138 (2019) 117–125.
19. Bo-qi Jia, Luo Xie, Xiao Cui, Li-jun Yang, and Qing-fei Fu. Linear stability of confined coaxial jets in the presence of gas velocity oscillations with heat and mass transfer. *Physics of Fluids*, 31 (2019) 092101.
20. Mukesh Kumar Awasthi, and Shivam Agarwal. Stability analysis between two concentric rotating cylinders with heat and mass transfer. *Heat Transfer: Asian Research*, 49(2) (2020) 971–983.
21. Weijia Qian, Yuzhen Lin, and Xin Hui. Effect of heat and mass transfer on the instability of an annular liquid sheet. *International Journal of Heat and Mass Transfer*, 143 (2019) 118499.
22. Mukesh Kumar Awasthi. Capillary instability of viscoelastic liquid film with heat and mass transfer. *Journal of Heat Transfer*, 142 (2020) 022108-1, doi: 10.1115/1.4045644
23. Shivam Agarwal, and Mukesh Kumar Awasthi. Pressure corrections for viscous potential flow analysis of Rayleigh-Taylor instability of swirling annular layer. *AIP Conference Proceedings*, 2220 (2020) 090005, https://doi.org/10.1063/5.0001673
24. Mukesh Kumar Awasthi. Kelvin-Helmholtz instability of viscoelastic liquid-viscous gas interface with heat and mass transfer. *International Journal of Thermal Sciences*, 161 (2021), doi.org/10.1016/j.ijthermalsci.2020.106710
25. Mukesh Kumar Awasthi, and Manu Devi. Temporal instability of swirling annular layer with mass transfer through porous media. *Special Topics & Reviews in Porous Media: An International Journal*, 11(1) (2020) 61–70.

# 3 Control Instruments of Regularized Problems Based on Mathematical Modeling of Structural Perturbations with Applications at the Nodes of 25-Bar Truss Systems

*Koumbe Mbock, Etoua Remy Magloire, and Ayissi Raoul Domingo*

**CONTENTS**

## 3.1 INTRODUCTION

In regularization, the use of $\mathcal{L}^1$-norm or $\|.\|_{\mathcal{L}^1(\Omega)}$ promotes solutions that are more sparse, and this norm comes from the difficulties of finding the optimal placement of discrete actuators (Stadler, 2009). The elliptic optimal control problems were studied for optimal heating where the control is the distribution source (Bergounioux, 1993a, 1993b; Bonnans, 1998; Tröltzsch, 2010; Casas and Tröltzsch 2012; Casas et al., 1998; 2012; 2014; 2017). Some convergence and regularization results were presented with sparsity constraints (Wachsmuth and Wachsmuth, 2011a, 2011b, 2013). In many applications, only perturbed measurements are available and every discretization of physical systems into integral or differential model equations introduces perturbations. The $\mathcal{L}^1(\Omega)$-regularizer is non-differentiable and it does not have a closed form solution in the context of optimal control problems. One possibility is to add additional $\mathcal{L}^2(\Omega)$ penalties or $\mathcal{L}^\infty(\Omega)$ bounds on the control. The penalization technique, using $\mathcal{L}^2$-norms, is understood as the Tikhonov regularization (Tikhonov and Arsenin, 1977) and the objective function is interpreted in the context of optimal control, as control costs with the positive regularization parameter $\alpha$. This regularization technique was developed recently to regularize ill-posed optimal control problems subject to elliptic partial differential equations (Engl et al., 1996; Tröltzsch, 2010; Wachsmuth and Wachsmuth, 2011a, 2011b; Wachsmuth, 2013; Pörner, 2018).

To recover some interesting behavior of the optimal control, it is sometimes necessary to compute a solution with a very small regularization parameter $\alpha$. But the Tikhonov regularization becomes increasingly ill-conditioned if $\alpha$ becomes very small. To obtain convergence without the requirement that $\alpha$ tends to zero, the proximal point method was introduced by Martinet (1970) and developed by many researchers (Rockafellar, 1976; Güler, 1991; Kaplan and Tichatschke, 1994, Hanke and Groetsch, 1998; Roting, 1999; Solodov and Svaiter, 2000; Reich and Sabach, 2010).

The iterative Bregman regularization method is another suitable regularization technique, based on the iterative Tikhonov regularization technique (Bregman, 1967). This method was developed to regularize ill-posed physical problems by replacing the Hilbert space norm with the Bregman distance (Eckstein, 1993; 1998; Eckstein and Bertsekas, 1992; Burger et al., 2007; Burger 2016). Recently, the convergence and stability of the regularization methods have been achieved for ill-posed optimal control by Pörner (2018). He extended the regularity conditions used for Bregman iterations to solve ill-posed optimal control problems, subject to control constraints. He discretized the physical problems governed by Poisson's equation with homogeneous Dirichlet boundary conditions, and established convergence and stability results. He solved the ill-posed optimal control problem with state and control constraints by coupling the Tikhonov regularization with the augmented Lagrange method (Bergounioux, 1993a; Frick et al., 2011; Frick and Grasmair, 2012; Ito and Kunish, 1990; Karl and Wachsmuth, 2017; Kanzow et al., 2018).

In the book by Pörner (2018), ill-posed optimal control problems subject to Poisson's equations appear in a wide range of mathematical applications. The convergence analysis of the appropriate regularization techniques depends on the nature of the differential equations and the cause of ill-posedness. The compactness of the solution operator describing Poisson's equation is the mathematical cause of the ill-posedness. In the reconstruction space of infrastructures designed by Mbock et al. (2019a, 2019b), the performance of linear elastic model equations was tested with explicit consideration of structural perturbations. The efficient and inefficient structural perturbations were imposed for optimal mass finding of 25-bar truss systems and for optimal form finding of shallow footing, respectively. To reconstruct the initial state of structural perturbations, it is important to know how they have been introduced into the systems. Analytically, they are difficult to reconstruct and one way to overcome this barrier is to define them as the control instruments.

In this chapter, the infrastructures are ill-posed because the optimal control problems are ill-posed. The efficient and inefficient structural perturbations are modeled as the cause of the ill-posedness in Hilbert space. They are considered as the inverse solution of reconstruction problems of infrastructures. This inverse solution is then interpreted as a control instrument in order to formulate the regularized ill-posed optimal control problems with state and structural perturbations. This approach is illustrated with efficient structural perturbations imposed at the node elements of 25-bar truss systems in terms of loading conditions. The results show that the imposed structural perturbations are treated as the control instruments of regularized problems of a 25-bar truss with well-chosen regularization parameters.

For this, we outline the chapter in this section. The act of adding or removing the efficient or inefficient structural perturbations is modeled in Section 3.2 through the construction of a family of linear elastic model equations with explicit consideration of structural perturbations. The structural perturbations are defined by a continuous function in Hilbert space and the family of inverse problems of structural perturbations finding is interpreted as a family of constrained ill-posed optimal control problems. To overcome the ill-posedness in Section 3.3, we formulate the regularization problem for a family of ill-posed optimal control problems by promoting the Tikhonov regularization with sparsity constraints on the structural perturbations. In Section 3.4, this formulation is applied with real-world measurements on structural perturbations imposed at the node elements of a 25-bar truss system in terms of loading conditions. The results are interpreted to illustrate the control instruments of regularized problems of a 25-bar truss system with well-chosen regularization parameters. A discussion is opened in Section 3.5 about the difficulty of finding the physical and material representation of structural perturbations, either analytically or numerically. Section 3.6 summarizes this chapter, including some new perspectives.

## 3.2 FAMILY OF LINEAR ELASTIC PARTIAL DIFFERENTIAL EQUATIONS WITH EXPLICIT CONSIDERATION OF STRUCTURAL PERTURBATIONS

In this section, the problem of finding structural perturbations in the design of infrastructures is interpreted as an ill-posed problem of infrastructures. For this, the act of adding or removing the structural perturbations is modeled through a family of linear elastic model equations manipulated by the explicit perturbations. We consider that the infrastructures occupy a domain $\Omega$ of $N$-dimensional design space $\mathbb{R}^{N\geq 1}$ and the boundary of $\Omega$ is denoted by $\partial\Omega$. In general, the Sobolev space $\mathcal{H}^1(\Omega)$ and the Hilbert space $\mathcal{L}^2(\Omega)$ are considered to be the construction and reconstruction spaces, respectively, with $\mathcal{H}^1(\Omega)\subset\mathcal{L}^2(\Omega)$. For a given stress tensor $\sigma\in\mathcal{L}^2(\Omega)^N$, the structural perturbation field is denoted by $\tilde{\sigma}\in\mathcal{L}^2(\Omega)^N$, such that the difference $\sigma-\lambda\tilde{\sigma}=\sigma_0\in\mathcal{L}^2(\Omega)^{\mathrm{N}}$ is the original stress tensor, where $\lambda$ is a positive weighting factor according to the splitting introduced by Mbock et al. (2019a). We consider the family of linear elastic model equations with explicit consideration of structural perturbations:

$$\mathrm{div}\,\sigma_0=-\mathrm{f},\ \text{in}\ \Omega \tag{3.1}$$

$$\mathrm{u}=0,\ \text{in}\ \partial\Omega_{u0} \tag{3.2}$$

$$\mathrm{u}=\mathrm{u}_{\mathrm{ad}},\ \text{on}\ \partial\Omega_{u_{ad}} \tag{3.3}$$

$$\sigma_0\cdot\mathrm{n}_{\mathrm{s}}=\mathrm{g},\ \text{on}\ \partial\Omega_{f1} \tag{3.4}$$

$$\sigma_0\cdot\mathrm{n}_{\mathrm{s}}=0,\ \text{on}\ \partial\Omega_{f0} \tag{3.5}$$

The volume and surface force are $\mathrm{f}\in\mathcal{L}^2(\Omega)$ and $\mathrm{g}\in\mathcal{L}^2(\partial\Omega)$, respectively, the unit normal vector is $\mathrm{n}_{\mathrm{s}}$, and the overall boundary is $\partial\Omega=\partial\Omega_{u0}\cup\partial\Omega_{u_{ad}}\cup\partial\Omega_{f0}\cup\partial\Omega_{f1}$. In this perturbed model equation, the volume and surface force are corrupted by the structural perturbations in the design domain $\Omega$ and on the boundaries $\partial\Omega_{f0}$ and $\partial\Omega_{f1}$, respectively. The Sobolev space $\mathcal{H}_s=\{u\in\mathcal{H}^1(\Omega)\mid u=0 \text{ on } \partial\Omega_{u0} \text{ and } u=u_{ad} \text{ on } \partial\Omega_{u_{ad}}\}$ is the solution space and all the displacements $\mathrm{u}\in\mathcal{H}_s$. They cannot be found from the ill-posed variational formulation stated as:

$$\begin{aligned}\int_{\Omega}\mathrm{div}\,\sigma_0(\mathrm{x})\cdot\mathrm{v}(\mathrm{x})\,\mathrm{dx}&=-\int_{\Omega}\sigma_0(\mathrm{x})\cdot\nabla\mathrm{v}(\mathrm{x})\,\mathrm{dx}\\&+\int_{\partial\Omega}\sigma_0(x)v(x)\cdot n_s(x)\,ds,\ \forall v\in\mathcal{H}_s\end{aligned} \tag{3.6}$$

Those displacements cannot model all the deformations when the structural perturbation field $\tilde{\sigma}$ exists in design domain $\Omega$ and on the boundaries $\partial\Omega_{f0}$ and $\partial\Omega_{f1}$. The set of perturbed model equations, Eqs. (3.1)–(3.5), is rewritten as:

$$\mathrm{div}\,\sigma=-f+\lambda\,\mathrm{div}\,\tilde{\sigma},\ \text{in}\ \Omega \tag{3.7}$$

$$u=0,\ \text{on}\ \partial\Omega_{u0} \tag{3.8}$$

$$u=u_{ad},\ \text{on}\ \partial\Omega_{u_{ad}} \tag{3.9}$$

$$\sigma \cdot \mathrm{n_s} = \mathrm{g} + \lambda \widetilde{\sigma} \cdot \mathrm{n_s} \text{ on } \partial\Omega_{f1} \tag{3.10}$$

$$\sigma.\mathrm{n_s} = \lambda \tilde{\sigma}.\mathrm{n_s} \text{ on } \partial\Omega_{f0} \tag{3.11}$$

In the model equations, Eqs. (3.7), (3.10), and (3.11), the volume force f is corrupted by the term $\lambda$ div $\tilde{\sigma} \in \mathcal{L}^2(\Omega)$ and the surface force g is corrupted by the term $\lambda \widetilde{\sigma}.\mathrm{n_s} \in \mathcal{L}^2(\partial\Omega_{f1})$. Also, we assume that the structural perturbations term $\lambda \widetilde{\sigma}.\mathrm{n_s}$ exists on the boundary $\partial\Omega_{f0}$ where the loading force does not act. The divergent div $\sigma = -f + \lambda$ div $\tilde{\sigma}$ cannot measure with accuracy how much the flow is expanding at a given point $\mathrm{p} = (x_1, x_2, x_3, \ldots, x_N) \in \Omega$. In particular, the construction space of infrastructures is the Sobolev space $\mathcal{H}_s$, whereas the reconstruction space of infrastructures is the Hilbert space $\mathcal{L}^2(\Omega)$ with $\mathcal{H}_s \subset \mathcal{L}^2(\Omega)$. In Eqs. (3.7), (3.10), and (3.11), the perturbation field $\tilde{\sigma}$ exists in the Hilbert spaces $\mathcal{L}^2(\Omega)$, $\mathcal{L}^2(\partial\Omega_{f0})$ and $\mathcal{L}^2(\partial\Omega_{f1})$, such that $\tilde{\sigma} = A\nabla\tilde{\mathrm{u}}$, and where $A^{\mathrm{N}\times\mathrm{N}}$ is a symmetric matrix and $\tilde{\mathrm{u}} \in \mathcal{L}^2(\Omega)$ represents the explicit perturbations function. The perturbed partial differential equations, Eqs. (3.7)–(3.11), are manipulated by the explicit perturbation function, and can be rewritten as:

$$\operatorname{div} \sigma = -f + (\lambda \operatorname{div} A\nabla\tilde{u}), \text{ in } \Omega \tag{3.12}$$

$$u = 0, \text{ on } \partial\Omega_{u0} \tag{3.13}$$

$$u = u_{ad}, \text{ on } \partial\Omega_{u_{ad}} \tag{3.14}$$

$$\sigma.\mathrm{n_s} = \mathrm{g} + (\lambda A\nabla\tilde{\mathrm{u}}.\mathrm{n_s}) \text{ on } \partial\Omega_{f1} \tag{3.15}$$

$$\sigma.\mathrm{n_s} = (\lambda A\nabla\tilde{\mathrm{u}}.\mathrm{n_s}) \text{ on } \partial\Omega_{f0} \tag{3.16}$$

According to the experimental work of Mbock et al. (2019a), we consider the family of well-posed approximations:

a) In Eq. (3.12), the volume force **f** depends on how the explicit perturbations are manipulated in $\Omega$. We have div $\sigma \neq 0$ in $\Omega \Rightarrow \forall\tilde{u} \in \mathcal{L}^2(\Omega)$, $\exists f \in \mathcal{L}^2(\Omega)$, such that $-f + \lambda$ div $\tilde{\sigma} \approx -f(\tilde{u})$.
b) In Eq. (3.15) the surface force g depends on how the explicit perturbations are manipulated on $\partial\Omega_{f1}$. We have $\sigma.n_s \neq 0$ on $\partial\Omega_{f1} \Rightarrow \forall\tilde{u} \in \mathcal{L}^2(\Omega), \exists g \in \mathcal{L}^2(\partial\Omega_{f1})$, such that $g + \lambda\, \tilde{\sigma} \cdot n_s \approx g(\tilde{u})$
c) In Eq. (3.16), the surface force does not exist where the explicit perturbations act on $\partial\Omega_{f0}$. We have $\sigma.n_s \neq 0$ on $\partial\Omega_{f0} \Rightarrow \forall\tilde{u} \in \mathcal{L}^2(\Omega), \exists 0 \in \mathcal{L}^2(\partial\Omega_{f0})$, such that $\lambda\tilde{\sigma}.n_s \approx 0_{\mathcal{L}^2(\partial\Omega_{f0})}$

With the approximations mentioned above, the Dirichlet conditions (Eqs. (3.13) and (3.14)) are satisfied in the solution space $\mathcal{H}_s$, whereas the Neumann conditions, Eqs. (3.15) and (3.16), are imposed until we have the variational formulation, given by:

$$\int_{\Omega} A\nabla u \cdot \nabla v \, dx = \int_{\Omega} f(\tilde{u}) \cdot v \, dx + \int_{\partial\Omega_{f0}} g(\tilde{u}) \cdot v \, ds, \quad \forall v \in \mathcal{H}_s \tag{3.17}$$

Through the inequalities of Cauchy Schwarz and Pointcaré, the bilinear form $\int_{\Omega} A\nabla v \cdot \nabla v dx = a(u,v) \in \mathcal{H}_s(\Omega)$ admits a quadratic form denoted by $q(v) = \|v\|^2_{\mathcal{H}_s(\Omega)}$, when $u$ is replaced by v. According to the theory of Lax-Milgram, the unique solution of the variational problem, Eq. (3.17), is also the unique solution of the linear elastic model equations, stated as:

$$\text{div}\, \sigma = -f(\tilde{u}), \text{ in } \Omega \tag{3.18}$$

$$u = 0, \text{ on } \partial\Omega_{u0} \tag{3.19}$$

$$u = u_{ad}, \text{ on } \partial\Omega_{u_{ad}} \tag{3.20}$$

$$\sigma \cdot n_s = g(\tilde{u}), \text{ on } \partial\Omega_{f0} \tag{3.21}$$

$$\sigma \cdot n_s = 0, \text{ on } \partial\Omega_{f1} \tag{3.22}$$

The stability of this model solution depends continuously on the surface force $g(\tilde{u}) \in \mathcal{L}^2(\partial\Omega_{f0})$ and the volume force $f(\tilde{u}) \in \mathcal{L}^2(\Omega)$. In the Hilbert space, $\mathcal{H}_s$, the energy functional is expressed as:

$$J(v,\tilde{u}) = \frac{1}{2} q(v) - \left( \int_{\Omega} f(\tilde{u}) \cdot v \, dx + \int_{\partial\Omega_{f0}} g(\tilde{u}) \cdot v \, ds \right) \tag{3.23}$$

The displacements vector is obtained by minimizing J(v, $\tilde{u}$) and we have:

$$u = \arg \min_{v \in \mathcal{H}_s, \tilde{u} \in \mathcal{L}^2(\Omega)} J(v,\tilde{u}), \tag{3.24}$$

where $\tilde{u} \in \mathcal{L}^2(\Omega)$ represents the explicit perturbation function associated with the displacements $u \in \mathcal{H}_s$.

**Proposition 3.1:**

Let J(v, $\tilde{u}$) be the energy functional defined for $v \in \mathcal{H}_s(\Omega)$ and $\tilde{u} \in \mathcal{L}^2(\Omega)$. Let $u \in \mathcal{H}_s(\Omega)$ be the unique solution of the variational problem, Eq. (3.17).

1. If **u** is the unique solution of the boundary value problem (Eqs. (3.18)–(3.22)) obtained by solving the energy minimization problem (Eq. (3.23)), then u is also the unique approximated solution of the perturbed boundary values problem (Eqs. (3.1)–(3.5)).

2. If $f(\tilde{u}) \approx f$ and $g(\tilde{u}) \approx g$, then the unique solution of the perturbed boundary value problem (Eqs. (3.18)–(3.22) is also the unique solution of the linear elastic model:

$$\text{div } A\nabla u = -f, \text{ in } \Omega \tag{3.25}$$

$$u = 0, \text{ on } \partial\Omega_u \tag{3.26}$$

$$u = u_{ad}, \text{ on } \partial\Omega_{u_{ad}} \tag{3.27}$$

$$A\nabla u \cdot n_s = g, \text{ on } \partial\Omega_{f0} \tag{3.28}$$

$$A\nabla u \cdot n_s = 0, \text{ on } \partial\Omega_{f1} \tag{3.29}$$

In linear elasticity, the system of partial differential equations (Eqs. (3.25)–(3.29)) is often used without explicit consideration of the structural perturbations function $\tilde{u} \in \mathcal{L}^2(\Omega)$. The model equations (Eqs. (3.25)–(3.29)) can be rewritten with explicit perturbations as presented in Eqs. (3.18)–(3.22). The explicit perturbations are often hidden variables in engineering problems, and most of their declaration can be implicit in the design domain $\Omega$ and on the boundaries $\partial\Omega_{f0}$ and $\partial\Omega_{f1}$. This proposition is used to formulate the inverse problems of structural perturbations finding in linear elasticity.

### 3.2.1 Family of Constrained Ill-Posed Optimal Control Problems Due to Structural Perturbations

Since the structural perturbations are the cause of the ill-posedness, they are modeled through the explicit perturbation function $\tilde{u} \in \mathcal{L}^2(\Omega)$. Here, they are considered to be the inverse solution of reconstruction problems of infrastructures. The inverse problem of structural perturbation finding is then interpreted as a constrained ill-posed–optimal control problem.

Let $X = \mathcal{H}_s$ and $Y = \mathcal{L}^2(\Omega)$ be two Hilbert spaces. The linear operator $K : Y \rightarrow X$ is a given continuous operator, which describes the partial differential equation (PDE), Eq. (3.18). In the context of an inverse problem, we treat the PDE Eq. (3.18) as a linear operator equation through the linear finite element approximation (Larson and Bengzon, 2013), and it is stated as:

$$K\tilde{u} = F \tag{3.30}$$

The inverse solution $\tilde{u} \in Y$ is the cause of the ill-posedness. This operator solution includes the design and computational errors compared to the ill-posedness, based on the compactness of a given linear operator K (Pörner, 2018). The measured data $F \in X$ is the measured output generated by the system of PDEs, Eqs. (3.18)–(3.22), including the minimum displacements $u \in \mathcal{H}_s$ obtained from Eq. (3.24). The measured output can be expressed with the explicit perturbations of the operator. In the context of ill-posed optimal control problems, the inverse solution $\tilde{u}$ is called the unknown control and the measured output F is called the measured state. The linear operator equation Eq. (3.30) is ill-posed in the sense of Hadamard (1953) if the explicit perturbation function $\tilde{u}$ does not exist or if it is neither unique nor stable. Through the mathematical modeling of structural perturbations mentioned above, we assume that the explicit perturbation function $\tilde{u} \in \mathcal{L}^2(\Omega)$ exists in the reconstruction space of infrastructures. Numerically, the initial state of structural perturbations is difficult to reconstruct because their physical and material representation is complex. We consider that the problem of structural perturbation finding can be formulated as a PDE-constrained ill-posed optimal control problem, stated as:

$$\min_{\tilde{u} \in Y, u \in X} \frac{1}{2} \|F - F_d\|_X^2 \tag{3.31}$$

$$\text{subject to } K\tilde{u} = F, \text{ in } \Omega$$

$$F \leq F_c, \text{ in } \Omega$$

$$\tilde{u}_a \leq \tilde{u} \leq \tilde{u}_b, \text{ in } \Omega \tag{3.32}$$

The constrained optimization problem, Eqs. (3.31)–(3.32), consists of steering the control $\tilde{u}$ to have the measured state F as close as possible to the desired state, $F_d \in X$. In reconstruction space Y, the state and control constraints have to be imposed. The measured state, $F \in X$, of infrastructures cannot exceed the available capacity of productivity, $F_c \in X$. We use the set of feasible measured states denoted by $F_{ad} = \{F \in X \mid F \leq F_c \text{ in } \Omega\}$ in the reconstruction space. The desired state of infrastructures cannot be fixed when the explicit perturbations are added to or removed from the physical systems. These explicit perturbations do not occupy the whole design domain $\Omega$ and the set of admissible controls $\tilde{U}_{ad} = \{\tilde{u} \in \mathcal{L}^2(\Omega) \mid \tilde{u} \leq \tilde{u}_a \text{ on } \Omega_a \subset \Omega \text{ and } \tilde{u} \geq \tilde{u}_b \text{ on } \Omega_b \subset \Omega\}$ is viewed as the feasible set of explicit perturbation functions. In comparison with the tracking type functional of Pörner (2018), the desired state of infrastructures cannot be fixed in the case of structural perturbations. The desired state, $F_d \in X$, of problem Eqs. (3.31) and (3.32) is called the optimal design of infrastructures.

## 3.3 FAMILY OF REGULARIZED ILL-POSED OPTIMAL CONTROL PROBLEMS WITH STATE AND STRUCTURAL PERTURBATION CONSTRAINTS

This section deals with the construction of the family of regularized ill-posed optimal control problems with state and structural perturbation constraints. We assume that the efficient and inefficient structural perturbations can be added to or removed from the physical systems in order to find the optimal design $F_d \in X$. These types of physical systems can be controlled through the structural perturbations data in size, shape, and topology. In engineering problems, the structural perturbations do not always occupy all the design domain $\Omega$ and they can be made up of several holes. Our strategy is to choose the appropriate regularization parameters to avoid the deviation of the solutions of regularized problems. This requires the Tikhonov regularization with sparsity constraints being promoted respectively by the $\mathcal{L}^2$-norm and the $\mathcal{L}^1$-norm. With these requirements, the structural perturbations must be penalized, steered, and treated as the control instruments such that the measured state is as close as possible to the desired state. This can be formulated as the PDE-regularized ill-posed optimal control problem, stated as:

$$\min_{\tilde{u}\in Y, u\in X} J_{\alpha,\beta}\left(\tilde{u},u\right)=\frac{1}{2}\left\|F-F_d\right\|_X^2+\frac{\alpha}{2}\left\|\tilde{u}\right\|_{\mathcal{L}^2(\Omega)}^2 + \beta\left\|\tilde{u}\right\|_{\mathcal{L}^1(\Omega)}, \quad \text{for } \alpha>0 \text{ and } \beta\geq 0 \tag{3.33}$$

$$\text{subject to } \mathrm{K}\tilde{\mathrm{u}}=\mathrm{F}, \text{ in } \Omega$$

$$\mathrm{F}\leq \mathrm{F_c}, \text{ in } \Omega$$

$$\tilde{\mathrm{u}}_{\mathrm{a}}\leq\tilde{\mathrm{u}}\leq\tilde{\mathrm{u}}_{\mathrm{b}}, \text{ in } \Omega \tag{3.34}$$

where the parameters $\alpha > 0$ and $\beta \geq 0$ are well-chosen regularization parameters. In the absence of the exact structural perturbations, it is difficult to use only the construction and analysis of the solution methods of Pörner (2018) for solving problem Eqs. (3.33) and(3.34). According to the convergence results of the regularization techniques developed by Pörner, it is clear that the regularized problem Eqs. (3.33) and (3.34) may have a unique solution with inexact regularization parameters. To formulate the control instruments of regularized problems Eqs. (3.33) and (3.34) with well-chosen regularization parameters, we use the real-world measurements on the structural perturbations.

## 3.4 APPLICATIONS TO REAL-WORLD MEASUREMENTS: STRUCTURAL PERTURBATION MODELS IMPOSED AT THE NODES OF 25-BAR TRUSS SYSTEMS AND REGULARIZATION OF THE CONTROL INSTRUMENTS

In this section, the 25-bar truss system is ill-posed due to efficient structural perturbations imposed at the node elements, in terms of loading conditions. The infrastructure is ill-posed because the optimal control problem is ill-posed. In this context, the desired state is the optimal mass design of the 25-bar truss system and the structural perturbations are the control instruments. We show that the 25-bar truss system can be manipulated by the structural perturbations and that they are modeled in order to

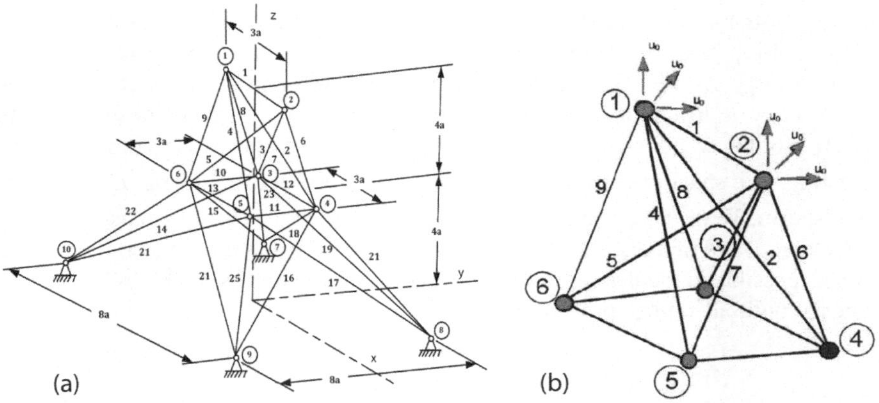


**FIGURE 3.1** (a) Twenty-five-bar truss system and (b) imposed loading conditions at the node elements {1}, {2}, {3}, {5}, and {6}, including the displacements at nodes {1} and {2} (from Mbock et al., 2019b).

**TABLE 3.1**
**Initial Structural Perturbation Data with Less Sparsity (Adapted from Mbock et al., 2019b)**

| Loading cases | Nodes | Imposed load on x-axis (kN) | Imposed load on y-axis (kN) | Imposed load on z-axis (kN) |
|---|---|---|---|---|
| 1 | 1 | 100 | 100 | –50 |
| 1 | 2 | 0 | 100 | –50 |
| 1 | 3 | 50 | 0 | 0 |
| 2 | 5 | 0 | 200 | –50 |
| 2 | 6 | 0 | –200 | 50 |

formulate the control instruments. Let us consider Figure 3.1 of a 25-bar truss system where the loading conditions and the displacements are imposed at the nodes.

In Figure 3.1b the structural perturbations are imposed in terms of efficient loading conditions at the five nodes, namely {1}, {2}, {3}, {5}, and {6}, as presented in Table 3.1.

In Table 3.2, the permissible displacements are fixed at the node elements {1} and {2}.

The truss system is manipulated by the explicit perturbations function $\tilde{u} \in \mathcal{L}^2(\Omega)$. Its behavior is modeled by the perturbed PDEs, Eqs. (3.18)–(3.22). We consider the mathematical perturbation model:

$$\operatorname{div} \sigma = -f(\tilde{u}), \text{ in } \Omega \tag{3.35}$$

**TABLE 3.2**
**Displacement Limit $u_0$ at Nodes {1}and {2} (from Mbock et al., 2019b)**

| Nodes | x-direction | y-direction | z-direction |
|---|---|---|---|
| 1 | ±3.5 | ±3.5 | ±3.5 |
| 2 | ±3.5 | ±3.5 | ±3.5 |

$$u = 0, \text{ on } \partial\Omega_{u0} = \partial\{\{7\},\{8\},\{9\},\{10\}\} \tag{3.36}$$

$$u = u_{ad}, \text{ on } \partial\Omega_{u_{ad}} = \partial\{\{1\},\{2\}\} \tag{3.37}$$

$$\sigma \cdot n_s = g_i(\tilde{u}), \text{ on } \partial\Omega_{f1} = \partial\{\{1\},\{2\},\{3\},\{5\},\{6\}\} \text{ with } i = 1,2,3,4,5 \tag{3.38}$$

$$\sigma \cdot n_s = 0, \text{ on } \partial\Omega_{f0} \tag{3.39}$$

We consider that the design parameters $(x_i)_{1 \le i \le 50}$ of the best profile of the 25-bar truss system are known, and the optimal mass is equal to 1963.908 kg, using the experimental data of Mbock et al. (2019b). After Proposition 3.1, this optimal mass is also the unique optimal mass obtained by solving the perturbed PDEs, Eqs. (3.35)–(3.39). Hence, we assume that the optimal mass design of the 25-bar truss system is called the desired state $F_d$ in the case of efficient structural perturbations and fixed in Table 3.3.

In the context of optimal control, the optimal mass design of the 25-bar truss system is called the desired state, and the structural perturbations are modeled as the control instruments. With the requirements of the Tikhonov regularization with sparsity constraints, we formulate the control instruments of the regularized problem of the 25-bar truss system with well-posed regularization parameters $\alpha > \beta > 0$. This is the PDE-regularized ill-posed optimal control problem, Eq. (3.33) and (3.34), in the form:

$$\min_{\tilde{u}\in Y, u\in X} J'_{\alpha,\beta}(\tilde{u}u) = \frac{1}{2}\|F - F_d\|_X^2 + \frac{\alpha}{2}\|\tilde{u}\|^2_{\mathcal{L}^2(\Omega)} + \beta\|\tilde{u}\|_{\mathcal{L}^1(\Omega)}, \text{ for } \alpha > \beta > 0 \tag{3.40}$$

$$\text{subject to } K\tilde{u} = F, \text{ in } \Omega$$
$$F \le F_c, \text{ in } \Omega$$
$$\tilde{u}_a \le \tilde{u} \le \tilde{u}_b, \text{ in } \Omega \tag{3.41}$$

**TABLE 3.3**
**Desired Values of the Optimal Design of the 25-Bar Truss System (from Mbock et al., 2019b)**

| Values of cross-sectional areas (cm²) | Values of stress elements (kN) | Stress constraints | Materials density (kN/m³) |
|---|---|---|---|
| $x_1 = 1.000$ | $x_{26} = -5.081$ | compression | 77.01 |
| $x_2 = 1.000$ | $x_{27} = -1.208$ | compression | 77.01 |
| $x_3 = 1.000$ | $x_{28} = 1.476$ | tension | 77.01 |
| $x_4 = 1.000$ | $x_{29} = 7.365$ | tension | 77.01 |
| $x_5 = 1.000$ | $x_{30} = -6.027$ | compression | 77.01 |
| $x_6 = 1.000$ | $x_{31} = -8.840$ | compression | 77.01 |
| $x_7 = 1.000$ | $x_{32} = -2.326$ | compression | 77.01 |
| $x_8 = 1.000$ | $x_{33} = -9.560$ | compression | 77.01 |
| $x_9 = 1.000$ | $x_{34} = 0.000$ | - | 77.01 |
| $x_{10} = 5.0905$ | $x_{35} = -9.428$ | compression | 77.01 |
| $x_{11} = 1.000$ | $x_{36} = -108.753$ | compression | 77.01 |
| $x_{12} = 1.000$ | $x_{37} = 11.605$ | tension | 77.01 |
| $x_{13} = 1.8753$ | $x_{38} = 1.432$ | tension | 77.01 |
| $x_{14} = 1.000$ | $x_{39} = -0.000$ | - | 77.01 |
| $x_{15} = 5.2082$ | $x_{40} = -1.470$ | compression | 77.01 |
| $x_{16} = 1.000$ | $x_{41} = 6.915$ | tension | 77.01 |
| $x_{17} = 9.6594$ | $x_{42} = -7.396$ | compression | 77.01 |
| $x_{18} = 1.000$ | $x_{43} = 0.000$ | - | 77.01 |
| $x_{19} = 1.000$ | $x_{44} = 40.065$ | tension | 77.01 |
| $x_{20} = 1.000$ | $x_{45} = 206.360$ | tension | 77.01 |
| $x_{21} = 1.000$ | $x_{46} = -79.786$ | compression | 77.01 |
| $x_{22} = 1.000$ | $x_{47} = -13.778$ | compression | 77.01 |
| $x_{23} = 1.000$ | $x_{48} = -111.226$ | compression | 77.01 |
| $x_{24} = 1.000$ | $x_{49} = 9.580$ | tension | 77.01 |
| $x_{25} = 3.7346$ | $x_{50} = 11.955$ | tension | 77.01 |

The regularization parameters are well chosen for many reasons. In Table 3.1, the structural perturbation data are not sparse in the whole design domain $\Omega$. We observe that the loading cases 1 and 2 are not zero in the large part of $\Omega$, and the $\mathcal{L}^1$-norm promotes sparsity with a percentage of 33.33%. The $\mathcal{L}^2$-norm promotes the Tikhonov regularization to avoid the deviation of the explicit perturbations, and the percentage of having fewer zeros is 66.66%. Also, in Table 3.1, the physical and material representation of the efficient structural perturbations may be unique, and their existence is difficult to establish analytically. Those structural perturbations are difficult to reconstruct numerically with the exact regularization parameters given by $\alpha > \beta > 0$, using only the regularization methods developed by Pörner (2018).

### 3.4.1 Interpretations of Results: Control Instruments of the Optimal Mass Design of 25-Bar Truss Systems with Loading Conditions Imposed at the Node Elements

The system of a 25-bar truss has to change in size, shape, and topology when the position of the efficient structural perturbations changes. For each change, we formulate the control instruments.

#### Example 3.1

The structural perturbation data of Table 3.1 are modified in Table 3.4.

The new permissible displacements are represented in Figure 3.2

In Table 3.5, the permissible displacements are fixed at the node elements {1},{2},{3}, and {5}.

The new truss system is manipulated by the structural perturbations data of Table 3.4, and its behavior is modeled by the perturbed PDEs, Eqs. (3.18)–(3.22). We have the mathematical perturbation model:

$$\operatorname{div} \sigma = -f(\tilde{u}), \text{ in } \Omega \tag{3.42}$$

$$u = 0, \text{ on } \partial\Omega_{u_0} = \partial\{\{7\},\{8\},\{9\},\{10\}\} \tag{3.43}$$

$$u = u_{ad}, \text{ on } \partial\Omega_{u_{ad}} = \partial\{\{1\},\{2\},\{3\},\{5\}\} \tag{3.44}$$

$$\sigma \cdot n_s = g_i(\tilde{u}), \text{ on } \partial\Omega_{f_1} = \partial\{\{1\},\{2\},\{3\},\{5\},\{6\}\} \text{ with } i = 1,2,3,4,5 \tag{3.45}$$

$$\sigma \cdot n_s = 0, \text{ on } \partial\Omega_{f_0} \tag{3.46}$$

**TABLE 3.4**
**Change in the Control Instruments of Table 3.1 without Sparsity**

| Loading cases | Nodes | Imposed load on x-axis (kN) | Imposed load on y-axis (kN) | Imposed load on z-axis (kN) |
|---|---|---|---|---|
| 1 | 1 | 100 | 100 | –50 |
| 1 | 2 | 50 | 100 | –50 |
| 1 | 3 | 50 | 100 | –50 |
| 2 | 5 | 50 | 200 | –50 |
| 2 | 6 | 50 | –200 | 50 |

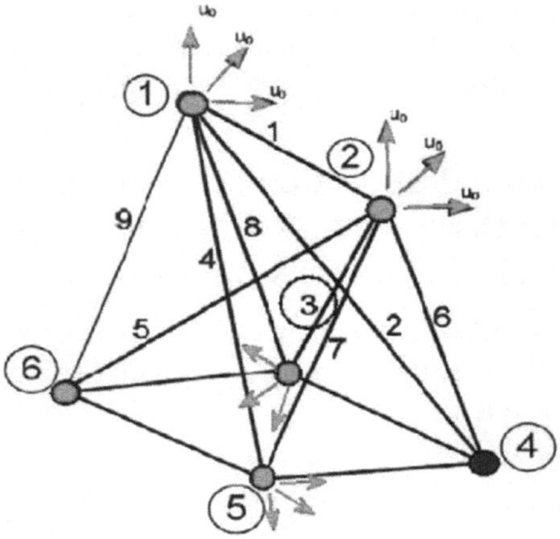


**FIGURE 3.2** Change in Figure 3.1b with the imposed loading conditions at nodes {1}, {2}, {3}, {5}, and {6}, including the displacements at the node elements {1}, {2}, {3}, and {5}.

**TABLE 3.5**
**Change in Table 3.2 with the Displacement Limit $u_0$ at the Nodes {1}, {2}, {3}, and {5}**

| Nodes | x-direction | y-direction | z-direction |
|---|---|---|---|
| 1 | ±3.5 | ±3.5 | ±3.5 |
| 2 | ±3.5 | ±3.5 | ±3.5 |
| 3 | ±3.5 | ±3.5 | ±3.5 |
| 5 | ±3.5 | ±3.5 | ±3.5 |

In the context of optimal control, the desired state $F_d$ is called the optimal mass design, and the structural perturbation data are modeled as the control instruments. In Table 3.4, the sparsity of structural perturbations is not mentioned and the percentage of having a zero is close to 0%, whereas the percentage of not having a zero is close to 100%. This indicates that the numerical value of the regularization parameter $\beta \geq 0$ will tend to zero and the parameter $\alpha > 0$ will be large. With the requirement of the Tikhonov regularization with sparsity constraints, we formulate the control instruments of the regularized problem of the 25-bar truss system with well-posed regularization parameters $\alpha > 0$ and $\beta \to 0$. This is the PDE-regularized ill-posed optimal control problem, Eqs. (3.33) and (3.34), in the form:

$$\begin{aligned} \min_{\tilde{u}\in Y, u\in X} J_{\alpha,\beta}(\tilde{u},u) &= \frac{1}{2}\|F-F_d\|_X^2 + \frac{\alpha}{2}\|\tilde{u}\|_{\mathcal{L}^2(\Omega)}^2 \\ &+ \beta\|\tilde{u}\|_{\mathcal{L}^1(\Omega)}, \text{ for } \alpha > 0 \text{ and } \beta \to 0 \end{aligned} \tag{3.47}$$

$$\text{subject to } K\tilde{u} = F, \text{ in } \Omega$$

$$F \le F_c, \text{ in } \Omega$$

$$\tilde{u}_a \le \tilde{u} \le \tilde{u}_b, \text{ in } \Omega \quad (3.48)$$

## Example 3.2

The structural perturbation data of Table 3.4 are modified in Table 3.6.

Also, the new permissible displacements are represented in Figure 3.3.

The permissible displacements are fixed at the node elements {1}, {2}, {3}, and {5}, as presented in Table 3.7.

**TABLE 3.6**
**Change in the Control Instruments of Table 3.4 with Too Great a Sparsity**

| Loading cases | Nodes | Imposed load on x-axis (kN) | Imposed load on y-axis (kN) | Imposed load on z-axis (kN) |
|---|---|---|---|---|
| 1 | 1 | 100 | 0 | 0 |
| 1 | 2 | 0 | 0 | –50 |
| 1 | 3 | 0 | 0 | 0 |
| 2 | 5 | 0 | 0 | 0 |
| 2 | 6 | 0 | –200 | 0 |

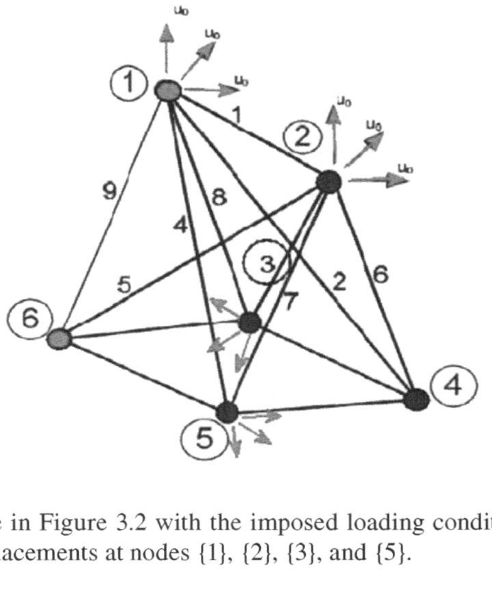


**FIGURE 3.3** Change in Figure 3.2 with the imposed loading conditions at nodes {1} and {6}, including the displacements at nodes {1}, {2}, {3}, and {5}.

**TABLE 3.7**
**Displacement Limit $u_0$ at the Nodes {1}, {2}, {3}, and {5} in Figure 3.3**

| Nodes | x-direction | y-direction | z-direction |
|---|---|---|---|
| 1 | ±3.5 | ±3.5 | ±3.5 |
| 2 | ±3.5 | ±3.5 | ±3.5 |
| 3 | ±3.5 | ±3.5 | ±3.5 |
| 5 | ±3.5 | ±3.5 | ±3.5 |

The new truss system is manipulated by the structural perturbation data of Table 3.6 and its behavior is modeled by the perturbed PDEs, Eqs. (3.18)–(3.22). We have the mathematical perturbation model:

$$\text{div}\,\sigma = -f(\tilde{u}),\ \text{in}\ \Omega \tag{3.49}$$

$$u = 0,\ \text{on}\ \partial\Omega_{u_0} = \partial\{\{7\},\{8\},\{9\},\{10\}\} \tag{3.50}$$

$$u = u_{ad},\ \text{on}\ \partial\Omega_{u_{ad}} = \partial\{\{1\},\{2\},\{3\},\{5\}\} \tag{3.51}$$

$$\sigma \cdot n_s = g_i(\tilde{u}),\ \text{on}\ \partial\Omega_{f_i} = \partial\{\{1\},\{6\}\}\ \text{with}\ i = 1,2 \tag{3.52}$$

$$\sigma \cdot n_s = 0,\ \text{on}\ \partial\Omega_{f_0} \tag{3.53}$$

In the context of optimal control, the desired state $F_d$ is called the optimal mass design and the structural perturbation data are modeled as the control instruments. In Table 3.6, the percentage of having many zeros is 80% and the percentage of having fewer zeros is around 20%. It indicates that the numerical value of the regularization parameter $\alpha > 0$ will be very small and the parameter $\beta > 0$ will be too large. With the requirement of the Tikhonov regularization with sparsity constraints, we formulate the control instruments of the regularized problem of the 25-bar truss with well-posed regularization parameters $\beta > \alpha > 0$. This is the PDE-regularized ill-posed optimal control problem, Eqs. (3.33) and (3.34) in the form:

$$\min_{\tilde{u}\in Y, u\in X} J_{\alpha,\beta}(\tilde{u},u) = \frac{1}{2}\|F - F_d\|_X^2 + \frac{\alpha}{2}\|\tilde{u}\|_{\mathcal{L}^2(\Omega)}^2 + \beta\|\tilde{u}\|_{\mathcal{L}^1(\Omega)},\ \text{for}\ \beta > \alpha > 0 \tag{3.54}$$

$$\text{subject to}\ K\tilde{u} = F,\ \text{in}\ \Omega$$

$$F \le F_c,\ \text{in}\ \Omega$$

$$\tilde{u}_a \le \tilde{u} \le \tilde{u}_b,\ \text{in}\ \Omega \tag{3.55}$$

### Example 3.3

The structural perturbation data of Table 3.6 are modified in Table 3.8.

Also, the new permissible displacements are represented in Figure 3.4.

The permissible displacements are fixed at the node elements {1},{2},{4}, and {6} as it is presented in Table 3.9.

The new truss system is manipulated by the structural perturbation data of Table 3.8, and its behavior is modeled by the perturbed PDEs, Eqs. (3.18)–(3.22). We have the mathematical perturbation model:

$$\operatorname{div}\sigma = -f(\tilde{u}),\ \text{ in }\ \Omega \tag{3.56}$$

$$u = 0,\ \text{ on }\ \partial\Omega_{u_0} = \partial\{\{7\},\{8\},\{9\},\{10\}\} \tag{3.57}$$

$$u = u_{ad},\ \text{ on }\ \partial\Omega_{u_{ad}} = \partial\{\{1\},\{2\},\{4\},\{6\}\} \tag{3.58}$$

$$\sigma \cdot n_s = 0,\ \text{ on }\ \partial\Omega_{f_1} \cup \partial\Omega_{f_0} \tag{3.59}$$

**TABLE 3.8**
**Change in the Control Instruments of Table 3.6 with Full Sparsity**

| Loading cases | Nodes | Imposed load on x-axis (kN) | Imposed load on y-axis (kN) | Imposed load on z-axis (kN) |
|---|---|---|---|---|
| 1 | 1 | 0 | 0 | 0 |
| 1 | 2 | 0 | 0 | 0 |
| 1 | 3 | 0 | 0 | 0 |
| 2 | 5 | 0 | 0 | 0 |
| 2 | 6 | 0 | 0 | 0 |

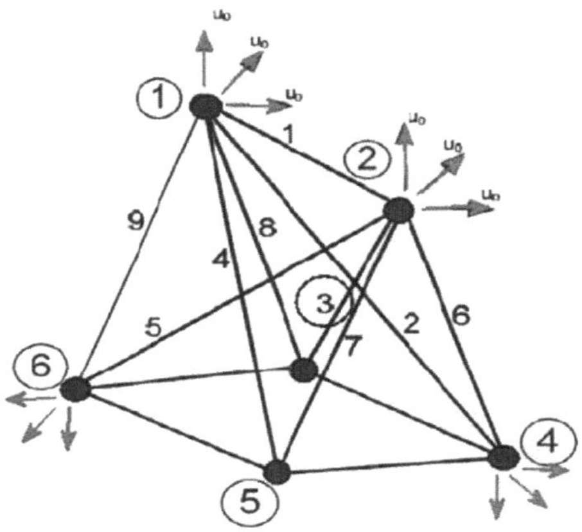


**FIGURE 3.4** Change in Figure 3.3 only with the displacements at nodes {1}, {2}, {4}, and {6}, and without loading conditions.

**TABLE 3.9**
**Displacement Limit $u_0$ at the Nodes {1}, {2}, {4},and {6} in Figure 3.4**

| Nodes | x-direction | y-direction | z-direction |
|---|---|---|---|
| 1 | ±3.5 | ±3.5 | ±3.5 |
| 2 | ±3.5 | ±3.5 | ±3.5 |
| 4 | ±3.5 | ±3.5 | ±3.5 |
| 6 | ±3.5 | ±3.5 | ±3.5 |

In the context of optimal control, the desired state $F_d$ is called the optimal mass design, and the structural perturbation data are modeled as the control instruments. In Table 3.8, the percentage of having many zeros is close to 100%, whereas the percentage of not having a zero is close to 0%. It indicates that the numerical value of the regularization parameter $\alpha > 0$ will tend to zero and the parameter $\beta > 0$ will be too large. With the requirements of the Tikhonov regularization with sparsity constraints, we formulate the control instruments of regularized problem of the 25-bar truss system with well-chosen regularization parameters $\beta > 0$ and $\alpha \to 0$. This is the PDE-regularized ill-posed optimal control problem, Eqs. (3.33) and (3.34), in the form:

$$\min_{\tilde{u} \in Y, u \in X} J_{\alpha,\beta}(\tilde{u},u) = \frac{1}{2}\|F - F_d\|_X^2 + \frac{\alpha}{2}\|\tilde{u}\|_{\mathcal{L}^2(\Omega)}^2 + \beta\|\tilde{u}\|_{\mathcal{L}^1(\Omega)}, \text{ for } \beta > 0 \text{ and } \alpha \to 0 \tag{3.60}$$

$$\text{subject to } K\tilde{u} = F, \text{ in } \Omega$$

$$F \le F_c, \text{ in } \Omega$$

$$\tilde{u}_a \le \tilde{u} \le \tilde{u}_b, \text{ in } \Omega \tag{3.61}$$

## 3.5 DISCUSSION

It is difficult to reconstruct the structural perturbation data of Tables 3.1, 3.4, 3.6, and 3.8 numerically. The regularization techniques developed by Pörner 2018 are not designed to find the physical and material representation of the structural perturbation data mentioned in this chapter. In many engineering problems, the structural perturbations are considered to be the deterministic cause of the ill-posedness, and they can include computational errors or the compactness of the linear operator. In the proposed approach, they are modeled first as the inverse solutions of the problem of structural perturbation finding, and then interpreted as the control instruments in Hilbert space. The Tikhonov regularization with sparsity constraints is required to construct the regularized problems with well-chosen regularization parameters. In the absence of exact regularization parameters, we apply the mathematical modeling of structural perturbations and the principles of regularization to avoid the deviation of structural perturbations. This strategy is applied to manipulate and control the systems of the 25-bar truss by the structural perturbations. With the knowledge of real-world measurements, the efficient structural perturbations are imposed at the node elements of 25-bar truss design. They are penalized and treated as the control instruments of regularized problems of a 25-bar truss with well-chosen regularization parameters $\alpha > 0$ and $\beta \geq 0$. The results of this approach show that the control instruments change when the perturbation models change, and the material and physical representation of structural perturbation data seem to be unique. It makes the construction of solution methods among the possible challenges for determining the values of regularization parameters numerically.

## 3.6 CONCLUSION

To formulate the control instruments of regularized problems in engineering, the act of adding or removing the structural perturbations has been modeled by a family of perturbed partial differential equations in linear elasticity. We saw that the structural perturbations are the cause of the ill-posedness and the inverse solution of the reconstruction problem. Those perturbations do not occupy the whole design domain. Their material and physical representation seem to be unique, with any, fewer, or many holes in the whole design domain. Those properties have been promoted to construct the family of regularized ill-posed optimal control problems with state and structural perturbation constraints. With real-world measurements, this approach has been applied to model the structural perturbations imposed at the nodes of the 25-bar truss system in terms of loading conditions. These explicit perturbations are penalized with the requirements of the Tikhonov regularization with sparsity constraints. The results have shown that the 25-bar truss system can be manipulated by efficient structural perturbations, and their initial state is difficult to reconstruct numerically. They are modeled and treated as the control instruments of regularized problems of the 25-bar truss system with well-chosen regularization parameters. These control instruments are formulated to have the measured state as close as possible to the optimal truss design. We saw that the control instruments change with well-chosen regularization parameters when the efficient structural perturbations change. This strategy can allow avoidance of inexact regularization parameters and regularization errors in the reconstruction problems of infrastructures. Therefore, it opens up new application fields and more theoretical perspectives for regularization techniques, with the challenge of proving the convergence of the adaptive parameter choice of regularization.

## CONFLICT OF INTEREST

The authors confirm that there is no conflict of interest to declare for this publication.

## ACKNOWLEDGMENTS

I thank Professor Etoua Remy Magloire and Professor Ayissi Raoul Domingo for this research on the control instruments in engineering problems at the National Advanced School Engineering of Yaounde (NASEY), Cameroon.

## REFERENCES

Bergounioux M. 1993a. Augmented Lagrangian method for distributed optimal control problems with state constraints. *Journal of Optimization Theory and Applications*, *78*(3), 493–521.

Bergounioux M. 1993b. On boundary state constrained control problems. *Numerical Functional Analysis and Optimization*, *14*(5–6), 515–543.

Bonnans J.F. 1998. Second-order analysis for control constrained optimal control problems of semilinear elliptic systems. *Applied Mathematics and Optimization*, 38(3), 303–325.

Bregman L.M. 1967. The relaxation method of finding the common point of convex sets and its application to the solution of problems in convex programming. *Ussr Computational Mathematics and Mathematical Physics*, *7*, 200–217.

Burger M. 2016. Bregman distances in inverse problems and partial differential equations. In *Advances in Mathematical Modeling, Optimization and Optimal Control*, volume 109 of Springer Optimization and Its Application (pp. 3–33). Springer, Cham.

Burger M., Resmerita E., & He L. 2007. Error estimation for Bregman iterations and inverse scale space methods in image restoration. *Computing*, *81*(2– 3), 109–135.

Casas E., & Tröltzsch F. 2012. Second order analysis for optimal control problems: improving results expected from abstract theory. *SIAM Journal of Optimization*, *22*(1), 261–279.

Casas E., & Tröltzsch F. 2014. Second-order and stability analysis for state constrained elliptic optimal control problems with sparse controls. *SIAM Journal of Optimization*, *52*(2), 1010–1033.

Casas E., Tröltzsch F., & Unger A. 1998. Second order sufficient optimality conditions for a class of elliptic control problems. In *Control and Partial Differential Equations (Marseille-Luminy, 1997)*, volume 4 of ESAIM Proceedings (pp. 285–300). Sociéte Mathématique Appliquées et Industrielles, Paris.

Casas E., Herzog R., & Wachsmuth G. 2012. Optimality conditions and error analysis of semilinear elliptic control problems with L-1 cost functional. *SIAM Journal of Optimization*, *22*(3), 795–820.

Casas E., Wachsmuth D., & Wachsmuth G. 2017. Sufficient second-order conditions for bang-bang control problems. *SIAM Journal on Control and Optimization*, *55*(5), 3066–3090.

Eckstein J. 1993. Nonlinear proximal point algorithms using Bregman functions, with applications to convex programming. *Mathematics of Operations Research*, *18*(1), 202–226.

Eckstein J. 1998. Approximate iterations in Bregman-function-based proximal algorithms. *Mathematical Programming*, *83*(1, Ser. A), 113–123.

Eckstein J., & Bertsekas D.P. 1992. On the Douglas-Rachford splitting method and the proximal point algorithm for maximal monotone operators. *Mathematical Programming*, *55*(3, Ser. A), 293–318.

Engl H.W., Hanke M., & Neubauer A. 1996. *Regularization of Inverse Problems*, volume 375 of Mathematics and Its Applications. Kluwer Academic Publishers Group, Dordrech.

Frick K., & Grasmair M. 2012. Regularization of linear ill-posed problems by the augmented Lagrangian method and variational inequalities. *Inverse Problems*, *28*(10), 104–105, 16.

Frick K., Lorenz D.A., & Resmerita E. 2011. Morozov's principle for the augmented Lagrangian method applied to linear inverse problems. *Multiscale Modeling and Simulation*, *9*(4), 1528–1548.
Güler O. 1991. On the convergence of the proximal point algorithm for convex minimization. *SIA M Journal on Control and Optimization*, *29*, 403–419.
Hadamard J. 1953. *Lectures on Cauchy's Problem in Linear Partial Differential Equations.* Dover Publications, New York.
Hanke M., & Groetsch C.W. 1998. Nonstationary iterated Tikhonov regularization. *Journal of Optimization Theory and Applications*, *98*(1), 37–53.
Ito K., & Kunisch K. 1990. The augmented Lagrangian method for equality and inequality constraints in Hilbert spaces. *Mathematical Programming*, *46*(3), 341–360.
Kanzow C., Steck D., & Wachsmuth D. 2018. An augmented Lagrangian method for optimization problems in Banach spaces. *SIAM Journal on Control and Optimization*, *56*(1), 272–291.
Kaplan A., & Tichatschke R. 2004. On inexact generalized proximal methods with a weakened error tolerance criterion. *Optimization*, *53*(1), 3–17.
Karl V., & Wachsmuth D. 2017. An augmented Lagrange method for elliptic state constrained optimal control problems. *Computational Optimization and Applications*, *45*(6), 598–603.
Larson, M.G. & Bengzon F. 2013. *The Finite Element Method: Theory, Implementation, and Applications*, volume 10. Springer, New York.
Martinet B. 1970. Brève communication. régularisation d'inéquations variationnelles par approximations successives. *ESAIM: Mathematical Modelling and Numerical Analysis: Modélisation Mathématique et Analyse Numérique*, *4*(R3), 154–158.
Mbock, K., Remy, M.E., Raoul D.A., Mpele M., & Richard, M.O. 2019a. On the optimal forms finding of shallow foundations made up of four foothills with explicit considerations of perturbations. *International Journal of Mathematical Engineering and Management Sciences*, *4*(3), 601–618.
Mbock, K., Remy. M.E., Seba L.M., Mpele M., & Richard, O.M. 2019b. Optimal Mass Design of 25 Bar Truss with Loading Conditions on Five Nodes Elements. *International Journal of Mathematical, Engineering and Management Sciences*, *4*(1), 1–16.
Pörner F. 2018. *Regularization Methods for Ill-Posed Optimal Control Problems.* Dissertation, Julius Maximilians-University Wuerzburg University Press.
Reich S., & Sabach S. 2010. Two strong convergence theorems for a proximal method in reflexive Banach spaces. *Numerical Functional Analysis and Optimization*, *31*(1–3), 22–44.
Rockafellar R.T. 1976. Augmented Lagrangians and applications of the proximal point algorithm in convex programming. *Mathematics of Operations Research*, *1*(2), 97–116.
Rotin S. 1999. *Konvergenz des Proximal-Punkt-Verfahrens für inkorrekt gestellte Optimalsteuerprobleme mit partiellen Differentialgleichungen.* PhD thesis, Universität Trier.
Solodov M.V., & Svaiter B.F. 2000. Forcing strong convergence of proximal point iterations in a Hilbert space. *Mathematical Programming*, *87*(1, Ser. A), 189–202.
Stadler G. 2009. Elliptic optimal control problems with L-control cost and applications for the placement of control devices. *Computational Optimization and Applications*, *44*, 159–181.
Tikhonov A.N., & Arsenin V.Y. 1977. *Solutions of Ill-Posed Problems.* Winston, New York.
Tröltzsch F. 2010. *Optimal Control of Partial Differential Equations*, volume 112 of Graduate Studies in Mathematics. American Mathematical Society, Providence, RI. Theory, methods and applications, Translated from the 2005 German original by Jürgen Sprekels.

Wachsmuth D. 2013. Adaptive regularization and discretization of bang-bang optimal control problems. *Electronic Transactions on Numerical Analysis*, *40*, 249–267.

Wachsmuth G., & Wachsmuth D. 2011a. Convergence and regularization results for optimal control problems with sparsity functional. *ESAIM: Control Optimization and Calculus* of *Variations*, *17*(3), 858–886.

Wachsmuth D., & Wachsmuth G. 2011b. Regularization error estimates and discrepancy principle for optimal control problems with inequality constraints. *Control and Cybernetics*, *40*(4), 1125–1158.

Wachsmuth D., & Wachsmuth G. 2013. Necessary conditions for convergence rates of regularizations of optimal control problems. In *System Modeling and Optimization*, volume 391 of IFIP Adv. Inf. Commun. Technol. (pp. 145–154). Springer, Heidelberg.

# 4 Numerical Simulation of Singularly Perturbed Differential Equation with Large Delay Using Exponential *B*-Spline Collocation Method

*Geeta Arora and Mandeep Kaur Vaid*

## CONTENTS



## 4.1 INTRODUCTION

Delay differential equations are the differential equations of particular interest to researchers for numerical simulation due to their expanded range of applications in biosciences, control theory, and many more fields. In recent years, these differential equations have been used in mathematical modeling of stability, systems control, diffusion in polymers, cancer immune systems, tumor studies, neurology, the pupil of the human eye, and many more fields. The small parameter known as the delay parameter arises in all systems which include feedback control. For example, the odd behavior of blood cells in a patient is due to a delay in growth, and this time delay appears as a result of a particular time period needed to sense the information and respond to it. A singular perturbed delay differential equation (SPDDE) is controlled by two small parameters known as the perturbation parameter and the delay parameter. The SPDDE shows eccentric behavior when the small parameter, which is multiplied by the highest order derivative term, tends toward zero.

Phenomena where these equations arise in real life in the biosciences include the study of diseases (Stein 1965, 1967; Longtin and Milton 1988; Bocharov and Rihan 2000; Rihan 2013), bifurcation (Derstine et al. 1982), the study of HIV infection (Nelson and Perelson 2002), the physiological control system (Mackey and Glass 1977), and the anti-cancer immune system (Wilkie and Hahnfeldt 2013). Much research work had been carried out to approximate the numerical solution of SPDDE by using various numerical techniques (Andargie and Reddy 2013; Nicaise and Xenophontos 2013; Swamy et al. 2015).

In the contemporary era, mathematical modeling of many real-life problems involves delay differential equations. Here, we briefly describe one such application to heating systems. The operation of thermal heating systems, such as solar heating systems, central heating systems, and so on, is based on simple thermal heating systems which are composed of a heater, storage, pipes, and a pump. The pump circulates the heat transfer fluid from the heater to the storage. The delay in heat transfer comes about due to the movement of fluid through pipes. The delaying effect of pipes plays an important role in the modeling of heater-storage systems in order to maximize the use of solar energy. A model developed for domestic hot water and for water heating in swimming pools (Kicsiny and Farkas 2012) took the delaying effect of pipes into consideration, and it was perceived that the use of solar energy was increased, compared with the models which disregarded the delaying effect of pipes. The use of the delay approach for complex heating systems, such as combustion systems, is supported by delay differential equations. Kicsiny (2014) developed one such model to elucidate heating systems with pipes. In his work, thermal engineering systems, involving heat and mass transfer, fluid mechanics, and the science of thermodynamics, were examined. The model was based on a white-box model. It considered both heat exchange and no heat exchange in the heating system, and the delay parameter depends upon the time and state of the heating system. This time delay was determined by the rate of flow of the fluid through the pump.

With this motivation, the exponential $B$-spline collocation method is used to solve singular perturbed differential equations with the delay parameter. In this chapter, some recent methods are analyzed in Section 4.2. The considered equation for approximation is described in Section 4.3. In Section 4.4, the numerical scheme used is described, and the convergence analysis of the proposed scheme is described in Section 4.5. In Section 4.6, numerical computations are carried out on some examples to test the efficiency of the scheme, followed by discussion and conclusions in Section 4.7.

## 4.2 ANALYSIS OF RECENT NUMERICAL WORK CARRIED OUT ON SPDDE

Subburayan (2018) presented an almost second-order convergent hybrid scheme to solve a second-order SPDDE with a discontinuous convention coefficient and a source function (term). The equation, which was considered for numerical treatment, was:

$$-\varepsilon y''(x) + a(x)\, y'(x) + b(x)\, y(x-1) = f(x)\ ,\ x \epsilon \Omega^{+}$$

subject to the condition $y(x) = \varnothing(x)$ for $x \epsilon [-1,0]$ and $y(2) = l$.

where $a(x)$ and $f(x)$ are discontinuous functions in interval [0,2]

Cengizci (2017) studied SPDDE of the second order by using an asymptotic numerical hybrid method. In that work, an SPDDE with a delay in the conventional term was considered as:

$$\varepsilon y''(x) + P(x) y'(x-\delta) + Q(x) y(x) = r(x)$$

with boundary condition $y(x) = \varnothing(x)$ for $-\delta \le x \le 0$ and $y(1) = \gamma$.

To solve this equation, an asymptotic method, which was well known as the successive complementary expansion method, was used to obtain a uniformly rational estimation method. The author compared the results with the exact solution, as well as with the existing method (File and Reddy 2014) to show the highly accurate numerical solution obtained by the asymptotic method presented.

File et al. (2017) solved a second-order SPDDE for which the delay existed in the reaction term by using a fourth-order uniformly convergent finite difference method. The considered SPDDE was:

$$\varepsilon^{2} y''(x) + a(x) y(x-\delta) + b(x) y(x) = f(x)\ ,$$

such that $y(x) = \varnothing(x)$ for $-\delta \le x \le 0$ and $y(1) = \beta$.

The proposed scheme was highly accurate as compared with the earlier methods (Swamy 2014; Soujanya et al. 2016) when tested with numerical examples.

Cimen (2017) presented a paper in which the author proposed a prior estimation of SPDDE with the delay parameter in the convention term by considering the following equation:

$$\varepsilon y''(x) + a(x) y'(x) + b(x) y'(x-r) + c(x) y(x-r) = f(x)\ , x \epsilon \Omega$$

subject to the conditions $y(x) = \varnothing(x)$ for $x \epsilon \Omega_0$ and $y(l) = A$.

In that work, the author elucidated the method to find the first- and second-order derivatives in the solution of SPDDE.

Kanth and Murali (2018) presented a numerical scheme to solve nonlinear SPDDEs. To convert the nonlinear SPDDE into a sequence of linear SPDDEs, the authors used the Quasilinearization technique to find the solution to the sequence of SPDDEs and implemented an exponentially fitted spline method. The considered equation was:

$$\varepsilon y''(x) = g(x, y, y'(x-\delta) \text{ on } (0,1)$$

such that $y(x) = \varnothing(x)$ for $-\delta \le x \le 0$ and $y(1) = \gamma$.

Recently, Sekar and Tamilselvan (2019) solved a class of SPDDE of the convection-diffusion type with integral boundary conditions by the first-order convergent finite difference scheme with a Shishkin mesh.

The considered problem was:

$$-\varepsilon y''(x) + a(x)y'(x) + b(x)y(x) + c(x)y(x-1) = f(x),\ x \in \Omega$$

under boundary conditions $y(x) = \varnothing(x)$ for $-1 \le x \le 0$ and $y(2) = l + \varepsilon \int_0^2 g(x)y(x)dx$

The authors tested the proposed scheme on four numerical examples and the maximum absolute pointwise error was presented as graphs.

## 4.3 CONSIDERED BOUNDARY VALUE PROBLEM

A second-order perturbed delay differential equation is given in the form

$$-\varepsilon y''(x) + a(x)y(x) + b(x)y(x-1) = f(x), \quad x \in \Omega^- \cup \Omega^+, \tag{4.1}$$

subject to the conditions

$$y(x) = \varnothing(x),\ x \in [-1,0],\ y(2) = \gamma \tag{4.2}$$

where $a(x) \ge \alpha_1 > \alpha > 0, \beta_0 \le b(x) \le \beta < 0$, and $\alpha_1 + \beta_0 > \eta > 0$. Also the functions $a(x)$, $b(x)$, and $f(x)$ are sufficiently smooth with respect to $\bar{\Omega}$.

This type of SPDDE, Eqs. (4.1)–(4.2), showcases the layer method at $x=0$ and at $x=2$, and the solution of such equations is well explained by Eloe et al. (2005). Furthermore, this type of boundary value problem occurs in control theory, where the problems involving signal transmission are exacerbated by the resulting time delays (Schmitt 1969). Some of the root causes liable for occurrence of these time delays have been ascertained, such as the medium of propagation of the signal and the excellence of the appliance used, whereas some of the causes are ambiguous (Kaufmann et al. 2012). Readers interested in such problems should refer to Elsgolts et al. (1964).

Taylor's series to second-order accuracy is used to handle the delay term:

$$y(x-1) = y(x) - y'(x) + y''(x)/2$$

By using the above relationship in Eq. (4.1), we obtain the equation:

$$P(x)y''(x) + Q(x)y'(x) + R(x)y(x) = f(x) \tag{4.3}$$

where $P(x) = \varepsilon - b(x)/2$, $Q(x) = -b(x)$ and $R(x) = b(x) + a(x)$

## 4.4 THE EXPONENTIAL CUBIC *B*-SPLINE COLLOCATION METHOD

The approximate solution of the equation is obtained by using an exponential *B*-spline basis function. To start with the procedure, the uniform mesh is considered in the domain $0 \leq x \leq 1$ with each sub-interval of length $1/N$, with $N$ as the number of partition points that results in the partition given by $\{0 = x_0 < x_1 < x_2 < \ldots\ldots. < x_N = l\}$. The approximate solution is thus given as:

$$y(x) = \sum_{i=-1}^{N+1} \alpha_i EB_i(x) \tag{4.4}$$

where $\alpha_i's$ are the unknown real coefficients and $EB_i(x)$'s are the exponential *B*-spline basis functions. The exponential *B*-splines, $EB_i$, at the partition points, together with the simulated points $x_{-3}, x_{-2}, x_{-1}, x_{N+1}, x_{N+2}, x_{N+3}$ outside the domain [*a*, *b*], can be defined as:

$$EB_i = \begin{cases} b_2((x_{i-2} - x) - \dfrac{\sinh(p(x_{i-2} - x))}{p}) for[x_{i-2}, x_{i-1}) \\ a_1 + b_1(x_i - x) + c_1 \exp(p(x_i - x)) + d_1 \exp(-p(x_i - x)) for[x_{i-1}, x_i) \\ a_1 + b_1(x - x_i) + c_1 \exp(p(x - x_i)) + d_1 \exp(-p(x - x_i)) for[x_i, x_{i+1}) \\ b_2((x - x_{i-2}) - \dfrac{\sinh(p(x - x_{i-2}))}{p}) for[x_{i+1}, x_{i+2}) \\ 0, otherwise \end{cases}$$

where

$$a_1 = \frac{pch}{phc - s}, b_1 = \frac{p}{2}\left[\frac{c(c-1)+s^2}{(phc-s)(1-c)}\right], b_2 = \frac{p}{2(phc-s)},$$

$$c_1 = \frac{1}{4}\left[\frac{\exp(-ph)(1-c)+s(\exp(-ph)-1)}{(phc-s)(1-c)}\right]$$

and $d_1 = \dfrac{1}{4}\left[\dfrac{\exp(ph)(c-1)+s(\exp(ph)-1)}{(phc-s)(1-c)}\right]$, $c = \cosh(ph)$, $s = \sinh(ph)$, $p$ is the free parameter.

So, $\{EB_{-1}(x), EB_0(x), EB_1(x), \ldots, EB_{N+1}(x)\}$ form the basis of functions defined over the interval [*a*, *b*]. Table 4.1 shows the values of $EB_i(x)$, $EB_i'(x)$ and $EB_i''(x)$ at the nodal points.

To know more about exponential *B*-spline, one can refer to Ersoy and Dag (2016).

**TABLE 4.1**
**Values of the Exponential Function and Its First- and Second-Order Derivatives at Knot Points**

| $x$ | $x_{i-2}$ | $x_{i-1}$ | $x_i$ | $x_{i+1}$ | $x_{i+2}$ |
|---|---|---|---|---|---|
| $EB_i(x)$ | 0 | $\frac{s-ph}{2(phc-s)}$ | 1 | $\frac{s-ph}{2(phc-s)}$ | 0 |
| $EB_i'(x)$ | 0 | $\frac{p(c-1)}{2(phc-s)}$ | 0 | $\frac{p(c-1)}{2(phc-s)}$ | 0 |
| $EB''_i(x)$ | 0 | $\frac{p^2s}{2(phc-s)}$ | $\frac{-p^2s}{phc-s}$ | $\frac{p^2s}{2(phc-s)}$ | 0 |

Using values of $EB_i(x), EB_i'(x), EB_i''(x)$ from Table 4.1 at nodal points, $y(x_i)$ and its derivatives are given as:

$$y(x_i) = m_1a_{i-1} + a_i + m_1a_{i+1},\ y'(x_i) = m_2a_{i+1} - m_2a_{i-1},\ y''(x_i) = m_3a_{i-1} - 2m_3a_i + m_3a_{i+1}$$

where $m_1 = \frac{s-ph}{2(phc-s)}$, $m_2 = \frac{p(c-1)}{2(phc-s)}$, $m_3 = \frac{p^2s}{2(phc-s)}$

Now, to apply the collocation technique, collocation points are selected in such a way that they concur with the nodal points and then, by substituting the values of $y_i, y_i'$ *and* $y_i''$ at nodal points in Eq. (4.4) we get a system of $N$+1 linear equations, with $N$+3 unspecified variables, as:

$$E_i^l\alpha_{i-1} + E_i\alpha_i + E_i^r\alpha_{i+1} = f_i, \qquad 0 \le i \le N \tag{4.5}$$

where $E_i^l = P(x)m_1 - Q(x)m_2 + R(x)m_3$, $E_i = P(x) - 2R(x)m_3$ and $E_i^r = P(x)m_1 + Q(x)m_2 + R(x)m_3$

The variables $\alpha_{-1}$ and $\alpha_{N+1}$ that exist when $i=0$ and $i=N$ are considered in Eq. (4.5). To eliminate these variables, boundary conditions $y(x_0) = \varnothing_0$ and $y(x_N) = \gamma$ are used that result in:

$$\alpha_{-1} = \frac{\varnothing_0 - \alpha_0 - m_1\alpha_1}{m_1} \text{ and } \alpha_{N+1} = \frac{\gamma - \alpha_N - m_1\alpha_{N-1}}{m_1}$$

Substituting these values in Eq. (4.5) for $i=0$ and $i=N$, we get:

$$\alpha_0\left(E_0 - \frac{1}{m_1}E_0^l\right) + \alpha_1\left(E_0^r - E_0^l\right) = f_0 - \frac{\varnothing_0 E_0^l}{m_1} \tag{4.6}$$

and

$$\alpha_{N-1}\left(E_N{}^{l}-E_N{}^{r}\right)+\alpha_N\left(E_N-\frac{1}{m_1}E_N{}^{r}\right)=f_N-\frac{\gamma E_N{}^{r}}{m_1} \tag{4.7}$$

This generates $N$+1 equations for $N$+1 variables that can be written as A $\alpha$=B where $\alpha=\alpha_0\ ,\alpha_1,\alpha_2,\ldots\ldots\ldots,\alpha_N$.

The tridiagonal matrix A is given by

$$\begin{bmatrix} E_0 & E_0^r & \cdots & \cdots & \cdots & \cdots & \cdots & 0 \\ E_1^l & E_1 & E_1^r & \cdots & \cdots & \cdots & \cdots & 0 \\ \vdots & \vdots & \vdots & \vdots & \vdots & \vdots & \vdots & \vdots \\ \vdots & \vdots & \vdots & \vdots & \vdots & \vdots & \vdots & \vdots \\ 0 & \cdots & 0 & E_i^l & E_i & E_i^r & 0 & \cdots \\ \vdots & \vdots & \vdots & \vdots & \vdots & \vdots & \vdots & \vdots \\ \vdots & \vdots & \vdots & \vdots & \vdots & \vdots & \vdots & \vdots \\ 0 & \cdots & \cdots & \cdots & \cdots & E_{N-1}^l & E_{N-1} & E_{N-1}^r \\ 0 & \cdots & \cdots & \cdots & \cdots & \cdots & E_N & E_N^r \end{bmatrix}$$

and the right-hand side is a column matrix B, given by

$$\begin{bmatrix} f(x_0)-E_0^l\left(\dfrac{\phi_0}{m_1}\right) \\ f(x_1) \\ f(x_2) \\ \vdots \\ \vdots \\ \vdots \\ f(x_{N-1}) \\ f(x_N)-\gamma(\dfrac{E_N^r}{m_1}) \end{bmatrix}$$

## 4.5 CONVERGENCE ANALYSIS

In this section, a procedure is described to discuss the convergence analysis of the method by truncation error. It is assumed that the function $y(x)$ is the function with continuous derivatives over the entire domain [0,1].

Now using the relation between $Y_N\left(x_i\right)$, $Y_N{}'\left(x_i\right)$, we get the following equation:

$$m_1Y_N{}'\left(x_{i-1}\right)+Y_N{}'\left(x_i\right)+m_1Y_N{}'\left(x_{i+1}\right)=m_2[y_N\left(x_{i+1}\right)+y_N\left(x_{i-1}\right)] \tag{4.8}$$

where $m_1$ and $m_2$ are the values as defined in Section 4.4.

By using the operator $E\left(y\left(x_i\right)\right) = y_N\left(x_{i+1}\right)$ in Eq. (4.8), we get

$$\left(m_1E^{-1} + 1 + m_1E\right)Y_N'\left(x_i\right) = m_2\left(E - E^{-1}\right)y_N\left(x_i\right)$$

Hence,

$$Y'_N\left(x_i\right) = \left[\frac{m_2\left(E - E^{-1}\right)}{\left(m_1E^{-1} + 1 + m_2E\right)}\right]y_N\left(x_i\right)$$

Substituting $E = e^{hD}$ and expanding it in powers of $hD$, we get

$$Y'_N\left(x_i\right) = \left[\frac{m_{2(}e^{hD} - e^{-hD})}{\left(m_1e^{-hD} + 1 + m_2e^{hD}\right)}\right]y_N\left(x_i\right)$$

$$= \frac{2m_2h}{\lambda}y_N'\left(x_i\right) + h^2\left[\frac{-2m_2\left(m_2 - m_1\right)}{\lambda^2}\right] + o\left(h^3\right)$$

where $\lambda = m_1 + m_2 + 1$

Therefore, the truncation error is: $h^2\left[\frac{-2m_2\left(m_2 - m_1\right)}{\lambda^2}\right] + o\left(h^3\right)$

Now the system of equations $AY=B$ is obtained, where $A = \left(t_{i,j}\right)$, $0 \le i, j \le N$, which is a tridiagonal matrix with

$$t_{i,i-1} = E_i^l,\ t_{i,i} = E_i \text{ and } t_{i,i+1} = E_i^r \text{ for } i = 1,2,3,4,\ldots,N-1$$

$$t_{0,0} = E_0,\ t_{0,1} = E_0^r,\ t_{N-1,N} = E_N \text{ and } t_{N,N} = E_N^r$$

with the local truncation error $T_i\left(h_i\right) = h^2\left[\frac{-2m_2\left(m_2 - m_1\right)}{\lambda^2}\right] + o\left(h^3\right)$
and $Y = (y_1, y_2, y_3, \ldots, y_{n-1})^t$

Also if $\overline{Y} = \left(\overline{y_1}, \overline{y_2}, \overline{y_3}, \ldots.\overline{y_{n-1}}\right)^t$ is the exact solution and the local truncation error is $T\left(h\right) = \left(T_1\left(h_1\right), T_2\left(h_2\right), T_3\left(h_3\right), \ldots\ldots\ldots. T_{n-1}\left(h_{n-1}\right)\right)^t$, then $A\overline{Y} - T\left(h\right) = C$, where $C$ is any constant, results in:

$$A\left(\overline{Y} - Y\right) = T\left(h\right)$$

The error equation can be written as

$$AE = T\left(h\right) \tag{4.9}$$

where $E = \bar{Y} - Y = (e_1, e_2, e_3, \ldots\ldots\ldots e_{n-1})^t$

Hence, from Eq. (4.9),

$$E = A^{-1} T(h) \tag{4.10}$$

The choice of $h$ is made sufficiently small, so that matrix A is irreducible and monotone. Thus $A^{-1}$ exists.

From the theory of matrices, it is known that $\sum_{i=0}^{n} \overline{t_{k,i}} S_i = 1$
where $\overline{t_{k,i}}$ is the $(k,i)$-th element of the $A^{-1}$.

Here we have $S_i = \sum_{j=1}^{n-1} t_{i,j} = h^0 B_i$,

$$\text{where } B_i = \begin{cases} P(x_i)(1+m_1) - Q(x_i)(m_2) + R(x_i)(-m_3)\, for\, i = 0 \\ P(x_i)(1+2m_1)\, for\, i = 1,2,3,\ldots\ldots, N-1 \\ P(x_i)(1+m_1) + Q(x_i)(m_2) - R(x_i)(-m_2)\, for\, i = N \end{cases}$$

Therefore, $\sum_{i=0}^{n} \overline{t_{k,i}} \le \frac{1}{\min(S_i)} \le \frac{1}{|B_i|}$

Thus, from Eq. (4.10), the element-wise error is $e_j = \sum_{i=0}^{n} \overline{t_{k,i}} T_i(h)$, for $j$=0,1,2,3,…,$N$, implies the result that $e_j \le \frac{kh^2}{|B_i|}$, where $k$ is a constant independent of $h$.

Therefore, $E = O(h^2)$

This concludes that our method is second-order convergent for the uniform mesh.

## 4.6 NUMERICAL EXAMPLES

Numerical solutions of some examples given by the presented scheme are discussed in this section in order to show the efficiency of the method. As the exact analytical solutions of the equations are not known, the double mesh principle is used to calculate the maximum absolute error and the uniform error defined as $D^N = \max \left| y_i^N - y_{2i}^{2N} \right|$ where $1 \le i \le N$.

### Example 4.1

Consider an SPDDE as $-\varepsilon y''(x) + 5y(x) - y(x-1) = 1$, under the conditions $y(x) = 1$, $x \in [-1, 0]$, $y(2) = 2$.

Table 4.2 shows the maximum error obtained for this example for $\varepsilon = 2^{-4}, 2^{-5}, \ldots, 2^{-23}$. The value of $D^N$ is compared with the results obtained from the reported method. It is observed from the results that, with an increase in the

**TABLE 4.2**
**Maximum Absolute Error Values Obtained for Example 4.1**

| $\varepsilon$ | $N=16$ | $N=32$ | $N=64$ | $N=128$ | $N=256$ | $N=512$ |
|---|---|---|---|---|---|---|
| $2^{-4}$ | 6.4760E-03 | 1.6120E-03 | 3.6600E-04 | 8.5000E-05 | 7.9000E-05 | 3.0800E-04 |
| $2^{-5}$ | 5.8860 E-03 | 1.4680E-03 | 3.3400E-04 | 7.4000E-05 | 7.6000E-05 | 3.0500E-04 |
| $2^{-6}$ | 6.1710 E-03 | 1.5370E-03 | 3.4900E-04 | 7.4000E-05 | 8.4000E-05 | 2.8100E-04 |
| $2^{-7}$ | 6.3200 E-03 | 1.5740E-03 | 3.6000E-04 | 7.7000E-05 | 8.4000E-05 | 3.1400E-04 |
| $2^{-8}$ | 6.3970E-03 | 1.5930E-03 | 3.6300E-04 | 8.0000E-05 | 6.9000E-05 | 3.1500E-04 |
| $2^{-9}$ | 6.4370E-03 | 1.6020E-03 | 3.6300E-04 | 7.9000E-05 | 5.7000E-05 | 2.8100E-04 |
| $2^{-10}$ | 6.4550 E-03 | 1.6060E-03 | 3.6700E-04 | 8.4000E-05 | 8.6000E-05 | 2.8700E-04 |
| $2^{-11}$ | 6.4650E-03 | 1.6100E-03 | 3.6600E-04 | 8.1000E-05 | 7.1000E-05 | 3.0700E-04 |
| $2^{-12}$ | 6.4710E-03 | 1.6100E-03 | 3.6600E-04 | 7.9000E-5 | 5.8000E-5 | 3.0600E-04 |
| $2^{-13}$ | 6.4730E-03 | 1.6110E-03 | 3.6700E-04 | 7.8000E-05 | 4.7000E-05 | 2.9200E-04 |
| $2^{-14}$ | 6.4740E-03 | 1.6110E-03 | 3.6800E-04 | 7.9000E-05 | 6.3000E-05 | 3.4200E-04 |
| $2^{-15}$ | 6.4750E-03 | 1.6110E-03 | 3.6800E-04 | 7.6000E-05 | 8.2000E-05 | 3.2900E-04 |
| $2^{-16}$ | 6.4760E-03 | 1.6110E-03 | 3.6600E-04 | 8.1000E-05 | 9.3000E-05 | 3.4400E-04 |
| $2^{-17}$ | 6.4750E-03 | 1.6120E-03 | 3.6600E-04 | 8.3000E-05 | 8.6000E-05 | 3.1700E-04 |
| $2^{-18}$ | 6.4750E-03 | 1.6110E-03 | 3.6700E-04 | 8.1000E-05 | 8.7000E-05 | 3.1400E-04 |
| $2^{-19}$ | 6.4750E-03 | 1.6120E-03 | 3.6800E-04 | 8.4000E-05 | 5.7000E-05 | 3.2600E-04 |
| $2^{-20}$ | 6.4760E-03 | 1.6100E-03 | 3.6900E-04 | 7.4000E-05 | 7.2000E-05 | 3.1300E-04 |
| $2^{-21}$ | 6.4750E-03 | 1.6120E-03 | 3.6500E-04 | 7.8000E-05 | 4.3000E-05 | 3.3900E-04 |
| $2^{-22}$ | 6.4760E-03 | 1.6110E-03 | 3.6700E-04 | 7.8000E-05 | 5.4000E-05 | 3.1400E-04 |
| $2^{-23}$ | 6.4760E-03 | 1.6110E-03 | 3.6700E-04 | 8.6000E-05 | 7.8000E-05 | 3.3500E-04 |
| $D^N$ by our method | 6.4760 E-03 | 1.6120E-03 | 3.6800E-04 | 8.6000E-05 | 9.3000E-05 | 3.4400E-04 |
| $D^N$ in Subburayan and Ramanujam (2013) | 1.2175 E-01 | 5.2206E-02 | 1.8447E-02 | 6.5158E-03 | 2.1589E-03 | 6.5625E-04 |

value of $N$ from 16 to 256, the maximum absolute error diminishes, whereas, for $N=512$, the obtained error again increases. Table 4.3 presents the maximum error obtained as shown for $\varepsilon = 0.01$, 0.001 and 0.0001, and it is found that, as the value of the perturbation parameter decreases, the absolute error increases. Table 4.4 represents the solution obtained by the presented scheme for some selected nodal points in the domain for $\varepsilon=0.01$ and for different values of $N$. Figure 4.1 presents the layer conduct of the approximate solution to this example for different values of the perturbation parameter $\varepsilon$.

### Example 4.2

Consider the equation given by $-\varepsilon y''(x)+(x+5)y(x)-y(x-1)=1$, subject to the conditions $y(x)=1,\ x\in[-1,0],\ y(2)=2$.

Figure 4.2 shows the layer behavior of the approximate solution to this example. Table 4.5 presents the maximum error obtained for this example for different values of $\varepsilon$ and $N$. It is clear that, as the value of $N$ increases from

**TABLE 4.3**
**Maximum Absolute Error Values Obtained from Example 4.1 for Different Values of $\varepsilon$**

| $\varepsilon$ | $N=16$ | $N=32$ | $N=64$ | $N=128$ | $N=256$ |
|---|---|---|---|---|---|
| 0.01 | 6.2770E-03 | 1.5640E-03 | 3.5500E-04 | 7.8000E-05 | 7.0000E-05 |
| 0.001 | 6.4550E-03 | 1.6070E-03 | 3.6600E-04 | 8.0000E-05 | 6.8000E-05 |
| 0.0001 | 6.4740E-03 | 1.6100E-03 | 3.6700E-04 | 8.1000E-05 | 6.8000E-05 |

**TABLE 4.4**
**Solution of Example 4.1 for $\varepsilon=0.01$**

| $x_i$ | $N=16$ | $N=32$ | $N=64$ | $N=128$ | $N=256$ | $N=512$ |
|---|---|---|---|---|---|---|
| 0.000 | 1.000000 | 1.000000 | 1.000000 | 1.000000 | 1.000000 | 1.000000 |
| 0.250 | 0.707339 | 0.707643 | 0.707723 | 0.707737 | 0.707745 | 0.707753 |
| 0.500 | 0.531473 | 0.531983 | 0.532117 | 0.532142 | 0.532154 | 0.532169 |
| 0.750 | 0.430268 | 0.431084 | 0.431296 | 0.431339 | 0.431354 | 0.431375 |
| 1.000 | 0.384337 | 0.385787 | 0.386158 | 0.386238 | 0.386261 | 0.386279 |
| 1.250 | 0.398841 | 0.401538 | 0.40222 | 0.402373 | 0.402412 | 0.40242 |
| 1.500 | 0.522984 | 0.527685 | 0.528865 | 0.529131 | 0.529195 | 0.529169 |
| 1.750 | 0.911869 | 0.918146 | 0.91971 | 0.920065 | 0.920143 | 0.920073 |
| 2.000 | 2.000000 | 2.000000 | 2.000000 | 2.000000 | 2.000000 | 2.000000 |

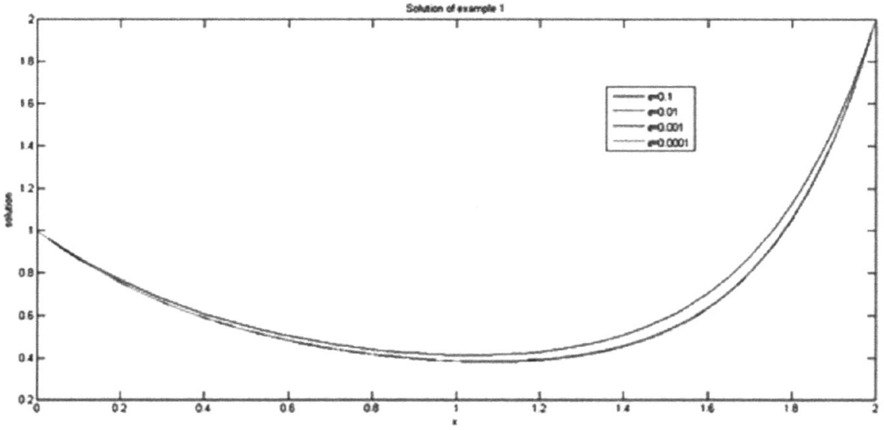


**FIGURE 4.1** Solution of Example 4.1 for different values of $\varepsilon$.

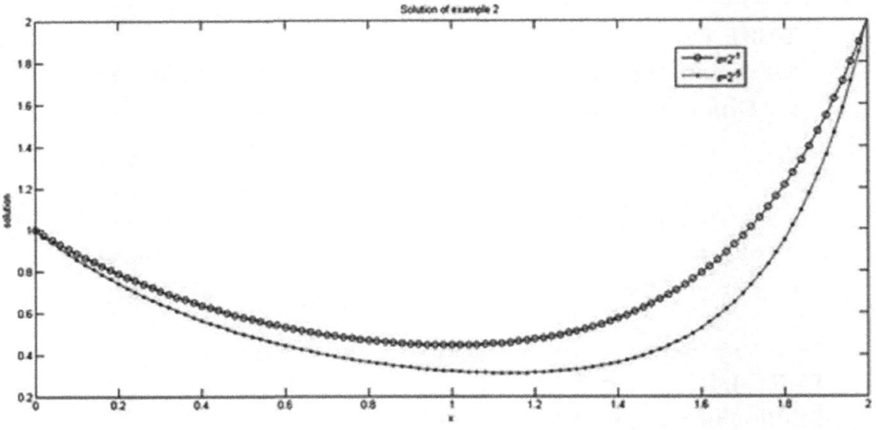


**FIGURE 4.2** Solution of Example 4.2 for different values of $\varepsilon$.

**TABLE 4.5**
**Maximum Absolute Error Values Obtained from Example 4.2 for Different Values of $\varepsilon$**

| $\varepsilon$ | $N=16$ | $N=32$ | $N=64$ | $N=128$ | $N=256$ | $N=512$ |
|---|---|---|---|---|---|---|
| $2^{-4}$ | 7.6480E-03 | 1.8930E-03 | 4.3900E-04 | 1.1700E-04 | 1.7000E-05 | 1.9000E-05 |
| $2^{-5}$ | 8.2800E-03 | 2.0440E-03 | 4.7600E-04 | 1.2600E-04 | 2.5000E-05 | 2.1000E-05 |
| $2^{-6}$ | 8.6270E-03 | 2.1260E-03 | 4.9700E-04 | 1.3200E-04 | 1.8000E-05 | 2.0000E-05 |
| $2^{-7}$ | 8.8090E-03 | 2.1700E-03 | 5.0600E-04 | 1.3100E-04 | 2.5000E-05 | 2.7000E-05 |
| $2^{-8}$ | 8.9020E-03 | 2.1930E-03 | 5.1300E-04 | 1.3500E-04 | 2.5000E-05 | 3.0000E-05 |
| $2^{-9}$ | 8.9490E-03 | 2.2030E-03 | 5.1600E-04 | 1.3600E-04 | 2.7000E-05 | 1.4000E-05 |
| $2^{-10}$ | 8.9730E-03 | 2.2100E-03 | 5.1600E-04 | 1.3600E-04 | 3.7000E-05 | 1.6000E-05 |
| $2^{-11}$ | 8.9850E-03 | 2.2120E-03 | 5.1700E-04 | 1.3600E-04 | 2.6000E-05 | 3.3000E-05 |
| $2^{-12}$ | 8.9910E-03 | 2.2140E-03 | 5.1800E-04 | 1.3600E-04 | 2.6000E-05 | 1.9000E-05 |
| $2^{-13}$ | 8.9940E-03 | 2.2130E-03 | 5.1900E-04 | 1.3700E-04 | 2.1000E-05 | 2.0000E-05 |
| $2^{-14}$ | 8.9960E-03 | 2.2150E-03 | 5.1800E-04 | 1.3400E-04 | 3.1000E-05 | 1.0000E-05 |
| $2^{-15}$ | 8.9970E-03 | 2.2140E-03 | 5.1700E-04 | 1.3800E-04 | 2.3000E-05 | 1.6000E-05 |
| $2^{-16}$ | 8.9970E-03 | 2.2150E-03 | 5.1800E-04 | 1.3800E-04 | 2.5000E-05 | 1.4000E-05 |
| $2^{-17}$ | 8.9970E-03 | 2.2160E-03 | 5.1700E-04 | 1.3800E-04 | 2.4000E-05 | 4.3000E-05 |
| $2^{-18}$ | 8.9970E-03 | 2.2150E-03 | 5.1600E-04 | 1.3800E-04 | 2.3000E-05 | 1.9000E-05 |
| $2^{-19}$ | 8.9970E-03 | 2.2150E-03 | 5.1700E-04 | 1.3700E-04 | 2.7000E-05 | 1.9000E-05 |
| $2^{-20}$ | 8.9970E-03 | 2.2150E-03 | 5.1800E-04 | 1.3600E-04 | 2.3000E-05 | 1.9000E-05 |
| $2^{-21}$ | 8.9970E-03 | 2.2140E-03 | 5.1900E-04 | 1.3700E-04 | 3.0000E-05 | 2.0000E-05 |
| $2^{-22}$ | 8.9970E-03 | 2.2160E-03 | 5.1800E-04 | 1.3400E-04 | 3.4000E-05 | 1.6000E-05 |
| $2^{-23}$ | 8.9970E-03 | 2.2150E-03 | 5.1900E-04 | 1.3500E-04 | 3.0000E-05 | 2.4000E-05 |
| $D^N$ by our method | 8.9970E-03 | 2.2160E-03 | 5.1900E-04 | 1.3800E-04 | 3.0000E-05 | 4.3000E-05 |
| $D^N$ in Subburayan and Ramanujam (2013) | 1.6717E-01 | 7.4650E-02 | 2.7140E-02 | 9.3384E-03 | 3.1724E-03 | 9.6871E-04 |

**TABLE 4.6**
**Maximum Absolute Error Values Obtained for Various Values of $\varepsilon$ from Example 4.2**

| $\varepsilon$ | $N=16$ | $N=32$ | $N=64$ | $N=128$ | $N=256$ |
|---|---|---|---|---|---|
| 0.01 | 8.7570E-03 | 2.1580E-03 | 5.0400E-04 | 1.2900E-04 | 2.6000E-05 |
| 0.001 | 8.9730E-03 | 2.2090E-03 | 5.1700E-04 | 1.3800E-04 | 2.7000E-05 |
| 0.0001 | 8.9950E-04 | 2.2140E-03 | 5.1900E-04 | 1.3700E-04 | 2.0000E-05 |

**TABLE 4.7**
**Approximate Solution of Example 4.2 for $\varepsilon=0.01$**

| $x_i$ | $N=16$ | $N=32$ | $N=64$ | $N=128$ | $N=256$ | $N=512$ |
|---|---|---|---|---|---|---|
| 0.000 | 1.000000 | 1.000000 | 1.000000 | 1.000000 | 1.000000 | 1.000000 |
| 0.250 | 0.687818 | 0.688021 | 0.688076 | 0.688082 | 0.68808 | 0.688086 |
| 0.500 | 0.493245 | 0.493629 | 0.493731 | 0.49375 | 0.493756 | 0.493753 |
| 0.750 | 0.375968 | 0.376655 | 0.376834 | 0.376871 | 0.376879 | 0.376885 |
| 1.000 | 0.314159 | 0.315527 | 0.315876 | 0.315954 | 0.315972 | 0.315979 |
| 1.250 | 0.308449 | 0.311315 | 0.312036 | 0.312203 | 0.312243 | 0.312251 |
| 1.500 | 0.405000 | 0.410684 | 0.412099 | 0.412428 | 0.412504 | 0.412527 |
| 1.750 | 0.777506 | 0.786263 | 0.788421 | 0.788925 | 0.789054 | 0.789072 |
| 2.000 | 2.000000 | 2.000000 | 2.000000 | 2.000000 | 2.000000 | 2.000000 |

16 to 256, the maximum absolute error decreases, but for $N=512$ a fluctuation in the error is observed. Table 4.6 shows the maximum error obtained as shown for $\varepsilon = 0.01, 0.001$ and $0.0001$. Table 4.7 presents the solution calculated for some nodal points in the domain.

### Example 4.3

We have considered SPDDE: $-\varepsilon y''(x)+2y(x)-2y(x-1)=1$, under the conditions $y(x)=1, x\in[-1,0], y(2)=2$.

Figure 4.3 shows that the behavior of the solution of this example is of a layer type. Tables 4.8 and 4.9 present the maximum error obtained for this example for various values of $\varepsilon$ and $N$. From the results obtained in Table 4.8, it can be deduced that the maximum absolute error decreases with increases in the total number of partition points for $\varepsilon = 2^{-5}, 2^{-7}, 2^{-13}$, but, for all other values of $\varepsilon$, the absolute error decreases from $N=16$ to 128 and then deviates. The calculated solution is presented in Table 4.10 for various values of $N$.

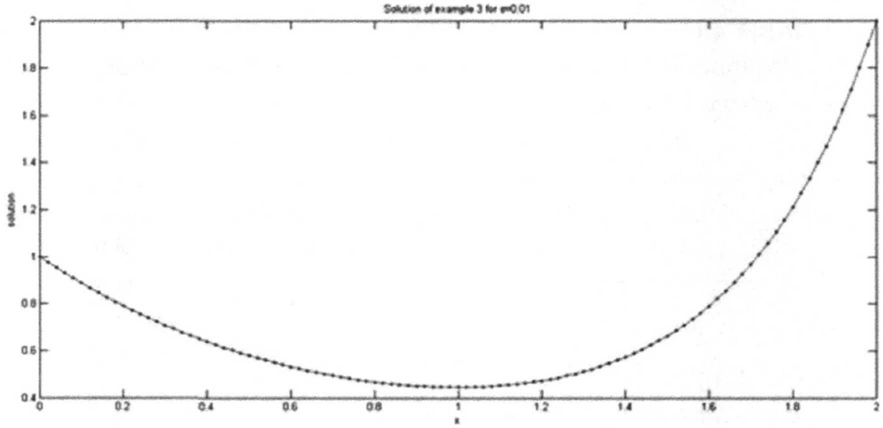


**FIGURE 4.3** Layer behavior of Example 4.3 for $\varepsilon = 0.01$.

**TABLE 4.8**
**Maximum Absolute Error Values Obtained from Example 4.3 for Different Values of $\varepsilon$**

| $\varepsilon$ | $N=16$ | $N=32$ | $N=64$ | $N=128$ | $N=256$ | $N=512$ |
|---|---|---|---|---|---|---|
| $2^{-4}$ | 8.0000E-04 | 2.2000E-04 | 4.0000E-05 | 3.0000E-05 | 7.0000E-05 | 7.0000E-05 |
| $2^{-5}$ | 8.9000E-04 | 2.4000E-04 | 4.0000E-05 | 4.0000E-05 | 4.0000E-05 | 2.8000E-04 |
| $2^{-6}$ | 9.3000E-04 | 2.5000E-04 | 5.0000E-05 | 1.0000E-05 | 4.0000E-05 | 1.0000E-04 |
| $2^{-7}$ | 9.5000E-04 | 2.6000E-04 | 4.0000E-05 | 3.0000E-05 | 2.0000E-05 | 1.1000E-04 |
| $2^{-8}$ | 9.6000E-04 | 2.6000E-04 | 5.0000E-05 | 3.0000E-05 | 5.0000E-05 | 1.1000E-04 |
| $2^{-9}$ | 9.6000E-04 | 2.6000E-04 | 3.0000E-05 | 4.0000E-05 | 4.0000E-05 | 9.0000E-05 |
| $2^{-10}$ | 9.6000E-04 | 2.6000E-04 | 4.0000E-05 | 2.0000E-05 | 3.0000E-05 | 9.0000E-05 |
| $2^{-11}$ | 9.7000E-04 | 2.6000E-04 | 4.0000E-05 | 2.0000E-05 | 3.0000E-05 | 2.3000E-04 |
| $2^{-12}$ | 9.7000E-04 | 2.6000E-04 | 5.0000E-05 | 2.0000E-05 | 5.0000E-05 | 2.3000E-04 |
| $2^{-13}$ | 9.7000E-04 | 2.6000E-04 | 3.0000E-05 | 3.0000E-05 | 2.0000E-05 | 1.2000E-04 |
| $2^{-14}$ | 9.7000E-04 | 2.6000E-04 | 4.0000E-05 | 3.0000E-05 | 6.0000E-05 | 3.0000E-04 |
| $2^{-15}$ | 9.7000E-04 | 2.6000E-04 | 3.0000E-05 | 5.0000E-05 | 6.0000E-05 | 6.0000E-05 |
| $2^{-16}$ | 9.7000E-04 | 2.6000E-04 | 5.0000E-05 | 2.0000E-05 | 5.0000E-05 | 1.7000E-04 |
| $2^{-17}$ | 9.7000E-04 | 2.5000E-04 | 4.0000E-05 | 6.0000E-05 | 7.0000E-05 | 2.9000E-04 |
| $2^{-18}$ | 9.8000E-04 | 2.6000E-04 | 3.0000E-05 | 2.0000E-05 | 9.0000E-05 | 7.0000E-05 |
| $2^{-19}$ | 9.7000E-04 | 2.7000E-04 | 4.0000E-05 | 1.0000E-05 | 1.2000E-04 | 3.6000E-04 |
| $2^{-20}$ | 9.7000E-04 | 2.6000E-04 | 3.0000E-05 | 1.0000E-05 | 3.0000E-05 | 1.5000E-04 |
| $2^{-21}$ | 9.7000E-04 | 2.6000E-04 | 4.0000E-05 | 2.0000E-05 | 3.0000E-05 | 1.1000E-04 |
| $2^{-22}$ | 9.7000E-04 | 2.6000E-04 | 4.0000E-05 | 1.0000E-05 | 2.0000E-05 | 1.4000E-04 |
| $2^{-23}$ | 9.7000E-04 | 2.6000E-04 | 4.0000E-05 | 1.0000E-05 | 2.0000E-05 | 5.0000E-05 |
| $D^N$ by our method | 9.7000E-04 | 2.6000E-04 | 5.0000E-05 | 6.0000E-05 | 1.2000E-04 | 3.6000E-04 |
| $D^N$ in Subburayan and Ramanujam (2013) | 7.0787E-02 | 3.0352E-02 | 1.0725E-02 | 3.7882E-03 | 1.2551E-03 | 3.8154E-04 |

**TABLE 4.9**
**Maximum Absolute Error Obtained for Various Values of $\varepsilon$ for Example 4.3**

| $\varepsilon$ | $N=16$ | $N=32$ | $N=64$ | $N=128$ | $N=256$ |
|---|---|---|---|---|---|
| 0.01 | 9.4000E-04 | 2.5000E-04 | 4.0000E-05 | 1.0000E-05 | 2.0000E-05 |
| 0.001 | 9.7000E-04 | 2.6000E-04 | 4.0000E-05 | 3.0000E-05 | 4.0000E-05 |
| 0.0001 | 9.6000E-04 | 2.7000E-04 | 4.0000E-05 | 1.0000E-05 | 3.0000E-05 |

**TABLE 4.10**
**Solution of Example 4.3 for $\varepsilon=0.01$**

| $x_i$ | $N=16$ | $N=32$ | $N=64$ | $N=128$ | $N=256$ | $N=512$ |
|---|---|---|---|---|---|---|
| 0.000 | 1.00000 | 1.00000 | 1.00000 | 1.00000 | 1.00000 | 1.00000 |
| 0.250 | 1.01231 | 1.01241 | 1.01244 | 1.01244 | 1.01244 | 1.01245 |
| 0.500 | 1.03255 | 1.03278 | 1.03284 | 1.03286 | 1.03285 | 1.03287 |
| 0.750 | 1.06582 | 1.06622 | 1.06633 | 1.06634 | 1.06635 | 1.06637 |
| 1.000 | 1.12049 | 1.1211 | 1.12126 | 1.12129 | 1.12129 | 1.12131 |
| 1.250 | 1.21036 | 1.21118 | 1.21139 | 1.21143 | 1.21144 | 1.21145 |
| 1.500 | 1.35809 | 1.35903 | 1.35928 | 1.35932 | 1.35932 | 1.35933 |
| 1.750 | 1.60090 | 1.60170 | 1.601910 | 1.60194 | 1.60195 | 1.60195 |
| 2.000 | 2.0000 | 2.0000 | 2.0000 | 2.0000 | 2.0000 | 2.0000 |

## 4.7 DISCUSSION AND CONCLUSIONS

In the present work, a boundary value problem for one type of SPDDEs is considered. To find an approximate solution for this type of problem, we have used an exponential $B$-spline collocation method. The method is shown to be a second-order convergent. We have applied the proposed scheme to three numerical examples and compared the results with those from the existing schemes, and it was found that our results were superior to the results reported in the literature. Some recent work, reported to solve the SPDDE, is also reviewed. The behavior of the approximate solution is presented in graphs for all the considered problems. From the results, it can be seen that the discussed method of exponential $B$-spline is capable of obtaining results of the required accuracy.

## REFERENCES

Andargie, A., and Y.N. Reddy. 2013. Parameter fitted scheme for singularly perturbed delay differential equations. *International Journal of Applied Science and Engineering* 11(4):361–373.

Bocharov, G.A., and F.A. Rihan. 2000. Numerical modelling in biosciences using delay differential equations. *Journal of Computational and Applied Mathematics* 125(1–2):183–199.

Cengizci, S. 2017. An asymptotic-numerical hybrid method for solving singularly perturbed linear delay differential equations. *International Journal of Differential Equations* 2017. www.hindawi.com/journals/ijde/2017/7269450

Cimen, E. 2017. A priori estimates for solution of singularly perturbed boundary value problem with delay in convection term. *Journal of Mathematical Analysis* 8(1): 202–211.

Derstine, M.W., H.M. Gibbs, F.A. Hopf, and D.L. Kaplan. 1982. Bifurcation gap in a hybrid optically bistable system. *Physical Review A* 26(6): 3720.

Eloe, P.W., Y.N. Raffoul, and C.C. Tisdell. 2005. Existence, uniqueness and constructive results for delay differential equations. *Electronic Journal of Differential Equations (EJDE)* [electronic only] 2005: 121–132.

Èl′sgol′c, L.E. 1964. *Qualitative Methods in Mathematical Analysis*. Vol. 12. American Mathematical Soc.

Ersoy, O., and I. Dag. 2016. The Exponential Cubic *B*-Spline Collocation Method for the Kuramoto-Sivashinsky Equation. *Filomat* 30(3):853–861.

File, G., and Y.N. Reddy. 2014. Terminal boundary-value technique for solving singularly perturbed delay differential equations. *Journal of Taibah University for Science* 8(3):289–300.

File, G., G. Gadisa, T. Aga, and Y.N. Reddy. 2017. Numerical solution of singularly perturbed delay reaction-diffusion equations with layer or oscillatory behaviour. *American Journal of Numerical Analysis* 5(1):1–10.

Kanth, A.R., and M.K.P. Murali. 2018. A numerical technique for solving nonlinear singularly perturbed delay differential equations. *Mathematical Modelling and Analysis* 23(1):64–78.

Kaufmann, P., P.L. Kaufmann, S.V.D. Pamboukian, and R.V. de Moraes. 2012. Signal transceiver transit times and propagation delay corrections for ranging and georeferencing applications. *Mathematical Problems in Engineering* 2012. https://pdfs.semanticscholar.org/9afb/50c8fd5a81584332cde9adc3a9770c9bf931.pdf?_ga=2.113708743.1128548923.1619864862-1038446754.1616057792

Kicsiny, R. 2014. New delay differential equation models for heating systems with pipes. *International Journal of Heat and Mass Transfer* 79:807–815.

Kicsiny, R., and I. Farkas. 2012. Improved differential control for solar heating systems. *Solar Energy* 86(11):3489–3498.

Longtin, A., and J.G. Milton. 1988. Complex oscillations in the human pupil light reflex with 'mixed' and delayed feedback. *Mathematical Biosciences* 90(1–2):183–199.

Mackey, M.C., and L. Glass. 1977. Oscillation and chaos in physiological control systems. *Science* 197(4300):287–289.

Nelson, P.W., and A.S. Perelson. 2002. Mathematical analysis of delay differential equation models of HIV-1 infection. *Mathematical Biosciences* 179(1):73–94.

Nicaise, S., and C. Xenophontos. 2013. Robust approximation of singularly perturbed delay differential equations by the hp finite element method. *Computational Methods in Applied Mathematics* 13(1):21–37.

Rihan, F.A. 2013. Delay Differential Equations in Biosciences: Parameter estimation and sensitivity analysis. In Recent Advances in Applied Mathematics and Computational Methods: Proceedings of the 2013 International Conference on Applied Mathematics and Computational Methods (Venice, Italy, September 2013) 2013:50–58.

Schmitt, K. 1969. On solutions of nonlinear differential equations with deviating arguments. *SIAM Journal on Applied Mathematics* 17(6):1171–1176.

Sekar, E., and A. Tamilselvan. 2019. Singularly perturbed delay differential equations of convection–diffusion type with integral boundary condition. *Journal of Applied Mathematics and Computing* 59(1–2):701–722.

Soujanya, G.B.S.L. and Y.N. Reddy. 2016. Computational method for singularly perturbed delay differential equations with layer or oscillatory behaviour. *Applied Mathematics and Information Sciences* 10(2):527–536.
Stein, R.B. 1965. A theoretical analysis of neuronal variability. *Biophysical Journal* 5(2):173–194.
Stein, R.B. 1967. Some models of neuronal variability. *Biophysical Journal* 7(1):37–68.
Subburayan, V. 2018. An hybrid initial value method for singularly perturbed delay differential equations with interior layers and weak boundary layer. *Ain Shams Engineering Journal* 9(4), 727–733.
Subburayan, V., and N. Ramanujam. 2013. An initial value technique for singularly perturbed reaction-diffusion problems with a negative shift. *Novi Sad Journal of Mathematics* 43(2), 67–80.
Swamy, D.K. 2014. Quantative analysis of delay differential equations with layer. *Advance Research and Innovations in Mechanical, Material Science, Industrial Engineering and Management* 145–150. http://bonfring.org/conference/papers/cjits_icarmmiem2014/mmiem-30.pdf
Swamy, D.K., K. Phaneendra, A.B. Babu, and Y.N. Reddy. 2015. Computational method for singularly perturbed delay differential equations with twin layers or oscillatory behaviour. *Ain Shams Engineering Journal* 6(1):391–398.
Wilkie, K.P., and P. Hahnfeldt. 2013. Mathematical models of immune-induced cancer dormancy and the emergence of immune evasion. *Interface Focus* 3(4):20130010.

# 5 Application of Differential Equations to Instability of Nanofluids

*Jyoti Sharma*

## CONTENTS



## 5.1 INTRODUCTION

The low thermal conductivity of fluids is a major factor in limiting improvements in the efficiency of heat transfer fluids, which are used in many industrial applications. To overcome this challenge, Maxwell proposed the possible incorporation of metallic particles into the heat transfer fluid to improve its thermal conductivity [1], but these particles settle out easily from the fluid due to their size and hence this approach was found not to be a feasible solution to the problem. Later, due to advances in nanotechnology, Choi visualized the concept of nanofluids to avoid size-related issues [2]. Research describing the considerable increase in heat transfer efficiency of fluids by the addition of nanoparticles has been reviewed in detail by Das and Choi [3] and Das et al. [4]. In addition to their industrial applications, nanoparticles are also being used in modern medicine, with copper oxide nanoparticles, having antibacterial properties, being utilized as antimicrobial agents. Copper nanoparticles have now also been found to be useful in drug delivery systems against infectious agents and cancer cells [5].

To explore mathematically the convective heat transfer phenomenon in nanofluids, Buongiorno derived partial differential equations to satisfy various conservation laws for fluids [6]. The convective instability (the Rayleigh-Bénard convection) in a horizontal layer was discussed rigorously by Chandrasekhar in his celebrated monograph on regular fluids [7]. Following [6] and [7], Tzou studied the thermal instability problem in nanofluids and found that the nanoparticles hasten the development of convection currents in the fluid [8, 9]. Nield and Kuznetsov investigated the same instability problem under different conditions on boundaries (free–free, rigid–free, and rigid–rigid) [10]. Gupta and co-workers [11] and Agarwal and colleagues [12] extended the nanofluid convection problems to study the effects of magnetic field and rotation, respectively, and established the stabilizing impact of both parameters on the nanofluid layer. Nield and Kuznetsov studied conservation equations to explore the dual impact of thermal and solutal diffusivities on the system [13]. Subsequently, their work was revisited, and numerical computations were made for an alumina–water nanofluid [14]. The equations for binary nanofluids were modified to include Lorentz [15] and Coriolis [16] terms and both parameters were found to delay the development of convection currents in nanofluids. Recently, Kumar and colleagues repeated the experiment with silver and selenium nanoparticles and confirmed the lower stability of nanofluids compared with regular fluids [17]. To explore porosity effects, the convection currents in a porous layer in the presence of nanoparticles were explored, along with the possibility of a chemical reaction [18]. Gupta and co-workers investigated the onset of convection currents in Casson nanofluids for an internally heated layer [19].

Nield and Kuznetsov revised the convective heat transfer model for nanofluids, and the nanoparticle flux across the boundaries of the layer was assumed to be zero [20]. The convection problems were revisited by a number of researchers [21–23], with the revised boundary conditions [20]. Subsequently, Sharma and colleagues considered the fact that the volume fraction is very small, and should be taken to be a constant at the basic state, to solve equations [24]. They established that the convection currents in the system are influenced by both the density and the conductivity of the nanoparticles. Recently, an extensive review on thermal and thermosolutal instability problems was published [25].

The present chapter considers two-dimensional partial differential equations for nanofluid flow, which include nanoscale effects. The horizontal layer is studied with respect to different initial conditions on nanoparticle volume fraction under Model 1, which varies it in the vertical direction, and Model 2, which assumes it to be so small that it is considered to be a constant. The basic state is disturbed slightly, and disturbed equations are solved analytically to obtain expressions for the Rayleigh number. In addition, the results are also discussed numerically and compared in detail for both cases in order to capture the essence of the process completely.

## 5.2 FORMULATION OF THE PROBLEM AND CONSERVATION EQUATIONS

A fluid layer of depth 'd' is considered in the $x$-direction between $y = 0$ and $y = d$ under a small temperature difference $(T_1 - T_0), T_1 > T_0$ as shown in Figure 5.1. The relevant conservation equations in two-dimensional form [6, 7, 10] are:

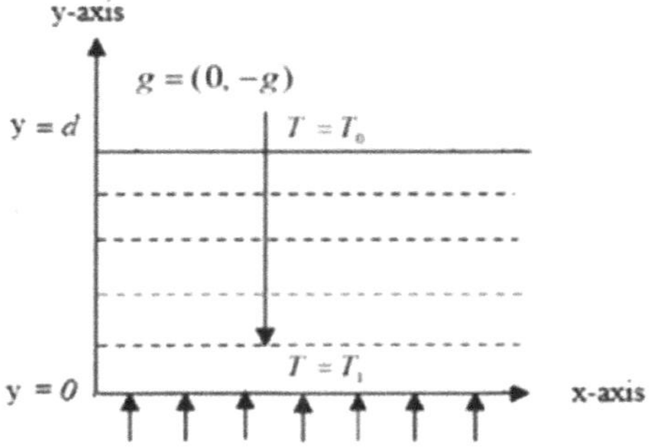


**FIGURE 5.1** Geometry of the problem.

Continuity equation (conservation of mass)

$$\frac{\partial u}{\partial x}+\frac{\partial v}{\partial y}=0, \tag{5.1}$$

where $\boldsymbol{v}=(u,v)$ is the velocity of the fluid.

Nanoparticle conservation equation

$$\frac{\partial \phi}{\partial t}+u\frac{\partial \phi}{\partial x}+v\frac{\partial \phi}{\partial y}=D_B\left(\frac{\partial^2\phi}{\partial x^2}+\frac{\partial^2\phi}{\partial y^2}\right)+\frac{D_T}{T_0}\left(\frac{\partial^2T}{\partial x^2}+\frac{\partial^2T}{\partial y^2}\right), \tag{5.2}$$

where $t$, $\phi$, $D_B$, and $D_T$ are time, nanoparticle volume fraction, and the coefficients of Brownian diffusion and thermophoretic diffusion, respectively.

Momentum equation in the $x$-direction

$$\rho_f\left(\frac{\partial u}{\partial t}+u\frac{\partial u}{\partial x}+v\frac{\partial u}{\partial y}\right)=-\frac{\partial p}{\partial x}+\mu\left(\frac{\partial^2u}{\partial x^2}+\frac{\partial^2u}{\partial y^2}\right), \tag{5.3}$$

where $\rho_f$, $p$, and $\mu$ are density, pressure, and viscosity of the fluid, respectively.

Momentum equation in the $y$-direction

$$\rho_f\left(\frac{\partial v}{\partial t}+u\frac{\partial v}{\partial x}+v\frac{\partial v}{\partial y}\right)=-\frac{\partial p}{\partial y}+\mu\left(\frac{\partial^2v}{\partial x^2}+\frac{\partial^2v}{\partial y^2}\right)$$
$$-\left[\phi\rho_p+(1-\phi)\rho_f(1-\beta_T(T-T_0)\right]g, \tag{5.4}$$

where $\rho_p$ and $\beta_T$ are the nanoparticle density and the volumetric coefficient of thermal expansion, respectively.

Thermal energy equation is

$$(\rho_f c)\left[\frac{\partial T}{\partial t}+u\frac{\partial T}{\partial x}+v\frac{\partial T}{\partial y}\right]=\kappa\left[\frac{\partial^2T}{\partial x^2}+\frac{\partial^2T}{\partial y^2}\right], \tag{5.5}$$

where $c$ and $\kappa$ are the specific heat and thermal conductivity of the fluid, respectively.

Equations (5.1)–(5.5) are partial differential equations, which are solved with initial and boundary conditions on a horizontal nanofluid layer to derive the expression for the thermal Rayleigh number under Model 1 and Model 2.

## 5.3 SOLUTION FOR MODEL 1: INITIALLY, VOLUME FRACTION VARIES IN THE VERTICAL DIRECTION

Let us assume the boundary conditions for nanoparticle volume fraction and temperature to be

$$\phi = \phi_1,\ T = T_1 \quad \text{at} \quad y = 0,$$

$$\phi = \phi_0,\ T = T_0 \quad \text{at} \quad y = d. \tag{5.6}$$

At the basic state, let the variables for the system vary as

$$v = 0,\ \phi = \phi_b(y),\ p = p_b(y),\ T = T_b(y). \tag{5.7}$$

Using Eqs. (5.1)–(5.6), we obtain

$$\frac{d^2\phi_b}{dy^2} = \frac{d^2T_b}{dy^2} = 0 \Rightarrow \phi_b(y) = \frac{(\phi_0 - \phi_1)}{d}y + \phi_1,\ T_b(y) = \frac{(T_0 - T_1)}{d}y + T_1. \tag{5.8}$$

The expression of pressure at the basic state is not required explicitly but can be obtained from Eq. (5.4). Let us assume that the system is disturbed slightly, and write

$$v = 0 + v',\ \phi = \phi_b + \phi', T = T_b + T', p = p_b + p'. \tag{5.9}$$

Using Eqs. (5.8) and (5.9) in Eqs. (5.1)–(5.5), we obtain

$$\frac{\partial u'}{\partial x} + \frac{\partial v'}{\partial y} = 0, \tag{5.10}$$

$$\frac{\partial \phi'}{\partial t} + \left(\frac{\phi_0 - \phi_1}{d}\right)v' = D_B\left(\frac{\partial^2\phi'}{\partial x^2} + \frac{\partial^2\phi'}{\partial y^2}\right) + \frac{D_T}{T_0}\left(\frac{\partial^2 T'}{\partial x^2} + \frac{\partial^2 T'}{\partial y^2}\right), \tag{5.11}$$

$$\rho_0\frac{\partial u'}{\partial t} = -\frac{\partial p'}{\partial x} + \mu\left(\frac{\partial^2 u'}{\partial x^2} + \frac{\partial^2 u'}{\partial y^2}\right), \tag{5.12}$$

$$\rho_0\frac{\partial v'}{\partial t} = -\frac{\partial p'}{\partial y} + \mu\left(\frac{\partial^2 v'}{\partial x^2} + \frac{\partial^2 v'}{\partial y^2}\right) - \left[\phi'(\rho_p - \rho_0) + \rho_0\beta_T T'\right]g, \tag{5.13}$$

$$(\rho_0 c)\left[\frac{\partial T'}{\partial t}+v'\frac{(T_0-T_1)}{d}\right]=\kappa\left[\frac{\partial^2 T'}{\partial x^2}+\frac{\partial^2 T'}{\partial y^2}\right]. \tag{5.14}$$

Partially differentiate Eq. (5.12) with respect to y and Eq. (5.13) with respect to $x$, and subtract to eliminate the pressure term, to obtain

$$\mu\left(\frac{\partial^2}{\partial x^2}+\frac{\partial^2}{\partial y^2}\right)^4 v'-\rho_0\frac{\partial}{\partial t}\left(\frac{\partial^2}{\partial x^2}+\frac{\partial^2}{\partial y^2}\right)^2 v'=\left[(\rho_p-\rho_0)\frac{\partial^2\phi'}{\partial x^2}+\rho_0\beta_T\frac{\partial^2 T'}{\partial x^2}\right]g. \tag{5.15}$$

Let us write disturbances as being exponential in time and periodic in the $x$-direction as

$$(v',\phi',T')=(V(y),\,\phi(y),\,T(y))e^{nt}\sin\alpha x, \tag{5.16}$$

where $n$ and $\alpha$ are growth rate and wave number, respectively.

The partial differential equations (5.11), (5.14), and (5.15) reduce to the ordinary and we obtain

$$\mu(D^2-\alpha^2)^2 V-n\rho_0(D^2-\alpha^2)V=\left[(\rho_p-\rho_f)(-\alpha^2)\phi-\rho_0\beta_T(-\alpha^2)T\right]g, \tag{5.17}$$

$$n\phi+\left(\frac{\phi_0-\phi_1}{d}\right)V=D_B(D^2-\alpha^2)\phi+\frac{D_T}{T_0}(D^2-\alpha^2)T, \tag{5.18}$$

$$(\rho_0 c)\left[nT+\frac{(T_0-T_1)}{d}V\right]=\kappa(D^2-\alpha^2)T \tag{5.19}$$

Substitute growth rate $n = 0$ for non-oscillatory motions, simplify Eqs. (5.17)–(5.19), and reduce to a non-dimensional form by substituting $\alpha = \alpha d$ and $y = yd$, to obtain

$$(D^2-\alpha^2)^3 V+Ra\alpha^2 V+\frac{(\rho_p-\rho_0)(T_0-T_1)D_T(\alpha^2)Vg}{T_0\alpha_f D_B}+\frac{(\phi_0-\phi_1)(\rho_p-\rho_0)d^3}{D_B\mu}\alpha^2 gV=0, \tag{5.20}$$

where the Rayleigh number $Ra=\rho_o g\beta d^3(T_0-T_1)/\mu\alpha_f$; $\alpha_f=k/\rho_0 c$

For the free–free boundaries [7]

$$V=D^2V=0 \text{ at } y=0,1. \tag{5.21}$$

Consider a trial solution to be $V = \sin\pi y$ to satisfy Eq. (5.21), and hence Eq. (5.20) gives

$$Ra=\frac{(\pi^2+\alpha^2)^3}{\alpha^2}-\left(\frac{D_T(T_0-T_1)}{\alpha_f T_0}+(\phi_0-\phi_1)\right)\frac{(\rho_p-\rho_0)\mathbf{g}\,d^3}{D_B\mu}. \tag{5.22}$$

For a regular fluid $(D_T = \phi_1 = \phi_0 = 0)$, Eq. (5.22) coincides with the expression given by Chandrasekhar [7]. It is worth mentioning that the nanoparticle parameters contribute to the reduction in the value of *Ra* (Eq. 5.22), with the destabilizing influence of nanoparticles being greater for a top-heavy ($\phi_0 > \phi_1$) case than for a bottom-heavy ($\phi_0 < \phi_1$) arrangement.

## 5.4 SOLUTION FOR MODEL 2: INITIALLY, VOLUME FRACTION REMAINS CONSTANT

The nanoparticle volume fraction in the nanofluid layer is so small that its influence would not vary in the vertical direction, which seems to be a more practical situation. Therefore, let us change the initial conditions for the nanoparticle volume fraction to solve the problem. This leads to a different set of equations and hence provides another significant expression for the Rayleigh number.

Let the nanoparticle volume fraction be constant ($\phi_0$) and all other variables be the same at the initial state. This assumption with regard to the initial solution satisfies Eqs. (5.1)–(5.5) without using any fixed conditions on the volume fraction at the boundaries of the layer. Adding small perturbations to the initial solution, the perturbed nanoparticle conservation equation takes the form

$$\frac{\partial \phi'}{\partial t} = D_B\left(\frac{\partial^2 \phi'}{\partial x^2} + \frac{\partial^2 \phi'}{\partial y^2}\right) + \frac{D_T}{T_0}\left(\frac{\partial^2 T'}{\partial x^2} + \frac{\partial^2 T'}{\partial y^2}\right), \tag{5.23}$$

with the remainder of the other equations (continuity, momentum, and energy) remaining the same. Solving Eqs. (5.14), (5.15), and (5.23) along similar lines gives the eigen value equations (for free–free boundaries) as

$$Ra\,\alpha^2 - \left(\pi^2 + \alpha^2\right)^3 + \frac{\left(\rho_p - \rho_0\right)\left(T_1 - T_0\right)D_T\left(\alpha^2\right)g}{T_0\,\alpha_f\,D_B} = 0. \tag{5.24}$$

Simplifying Eq. (5.24), we obtain

$$Ra = \frac{\left(\pi^2 + \alpha^2\right)^3}{\alpha^2} - \frac{D_T\left(\rho_p - \rho_0\right)(T_1 - T_0)g\,d^3}{D_B\mu\,\alpha_f T_0}. \tag{5.25}$$

It is interesting to note that Eq. (5.25) is independent of the nanoparticle volume fraction. Also, no boundary condition is imposed on the nanoparticle volume fraction in this case, meaning that the nanoparticles adjust themselves automatically to make the process happen, which seems to be a more realistic situation.

It is worth mentioning that, when we take $\phi_1 = \phi_0$, which is not a practical situation in general, Model 2 coincides with Model 1 in the most typical case. This means that, if we take the volume fraction constant at the initial state, nanoparticles may adjust themselves in a way that leads to a particular case of a constant volume fraction on the boundaries.

## 5.5 DISCUSSIONS AND COMPARATIVE STUDIES OF THE RESULTS

When we further simplify Eqs. (5.22) and (5.25), we obtain:

For Model 1

$$Ra = \frac{\left(\pi^2 + \alpha^2\right)^3}{\alpha^2} - \left(\frac{1}{2k + k_p} + (\phi_0 - \phi_1)\right)\left(\rho_p - \rho_0\right)A. \tag{5.26}$$

For Model 2

$$Ra = \frac{\left(\pi^2 + \alpha^2\right)^3}{\alpha^2} - \frac{\left(\rho_p - \rho_0\right)}{2k + k_p}A, \tag{5.27}$$

where '$A$' is independent of the physical properties of the nanoparticles [6]. The expression for the Rayleigh number under Model 1 includes one extra term relative to that obtained for Model 2. In other words, the instability of the system also depends on the difference in the particle volume fraction at the boundaries, in addition to other parameters, if the initial nanoparticles vary in the vertical direction, whereas this is not the case if the nanoparticles are considered to be constant at the basic state.

## 5.6 NUMERICAL RESULTS AND DISCUSSIONS

The expression of the thermal Rayleigh number is analyzed numerically to investigate the impact of different nanofluid parameters on convective heat transfer, with the help of the software Wolfram Mathematica. In Eqs. (5.26) and (5.27), within the permissible range of all the variables, variation in one parameter is considered, with all other variables being considered to be fixed, in order to plot the stability curves for different wave numbers.

Figure 5.2 depicts the stability curves for the thermal Rayleigh number *versus* wave number relationship for both models. Stability for Model 1 is observed to be less than that for Model 2, meaning that, when the nanoparticle volume fraction varies in the vertical direction at the initial state, the fluid layer is less stable than in the case where the nanoparticles are taken to be constant at the initial state. Furthermore, for lower values of wave numbers, the curves seem to coincide, and the critical wave number is the same for both curves.

Let us analyze the effect of differences in the nanoparticles at the boundaries of the nanofluid layer for all other variables considered to be fixed. Figure 5.3 shows that, as the nanoparticle volume fraction increases at the upper layer, the value of the thermal Rayleigh number decreases, and hence it hastens the convective instability of the layer. Figure 5.4 shows the destabilizing influence of nanoparticle density as the value for the Rayleigh number decreases in response to an increase in nanoparticle density, whereas, on the other hand, their conductivity stabilizes the layer (Figure 5.5).

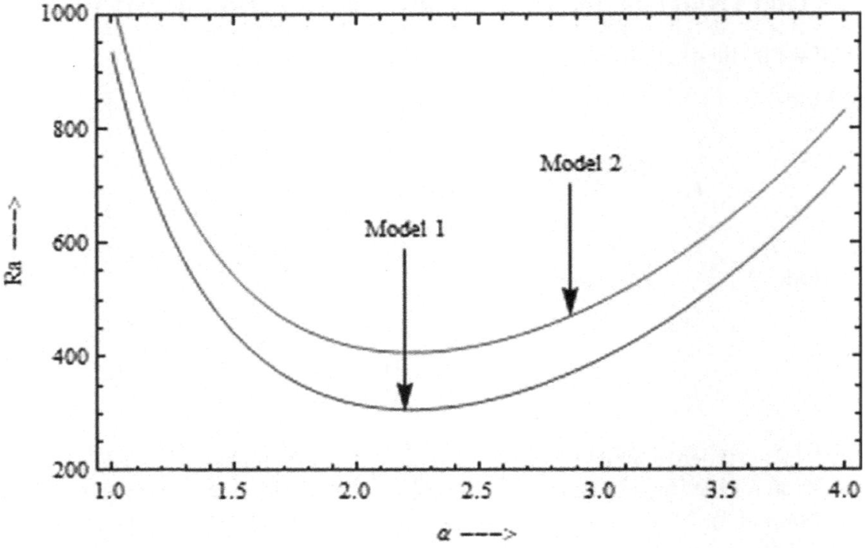


**FIGURE 5.2** Stability curves for Model 1 and Model 2.

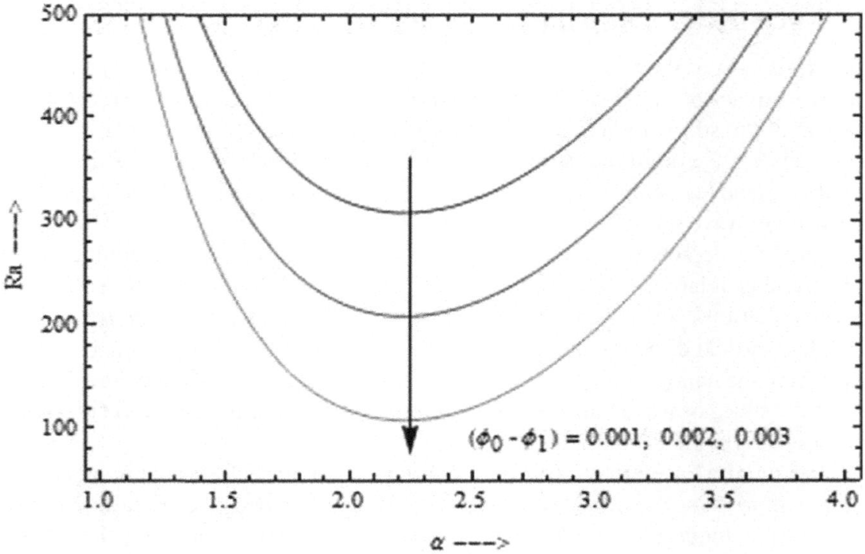


**FIGURE 5.3** Impact of nanoparticle volume fraction.

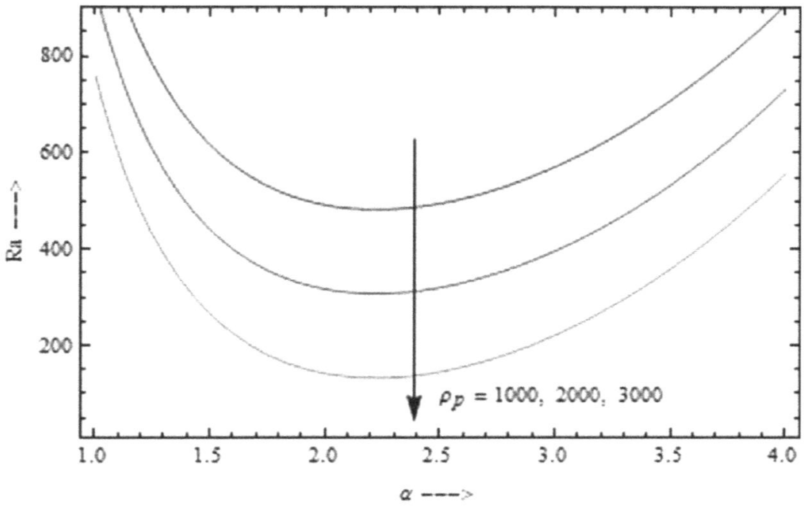


**FIGURE 5.4** Impact of nanoparticle density.

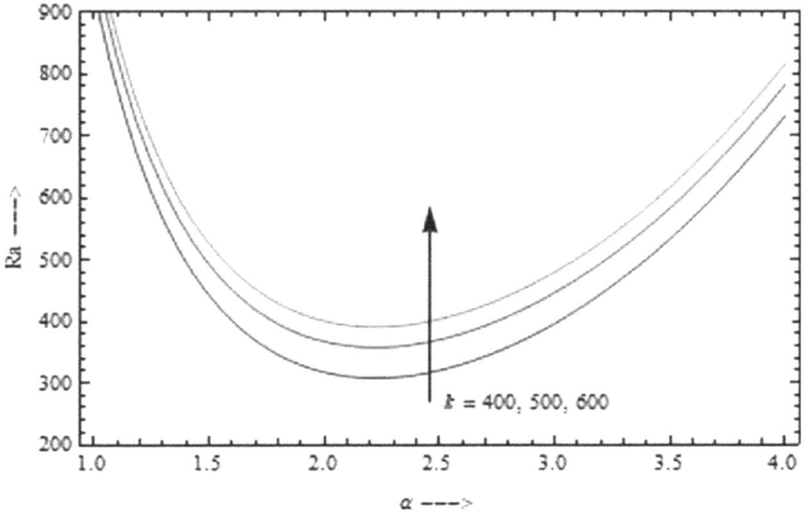


**FIGURE 5.5** Impact of nanoparticle conductivity.

**TABLE 5.1**
**Density and Conductivity of Nanoparticles [6, 24]**

| Physical Properties | Al | Ag | Au | Fe | $Al_2O_3$ | $SiO_2$ | CuO | $TiO_2$ |
|---|---|---|---|---|---|---|---|---|
| $\rho_p$ (kg/m$^3$) | 2700 | 10500 | 19320 | 7900 | 3970 | 2600 | 6510 | 4250 |
| $K_p$ (W/mK) | 237 | 429 | 314 | 80 | 40 | 10.4 | 18 | 8.9 |

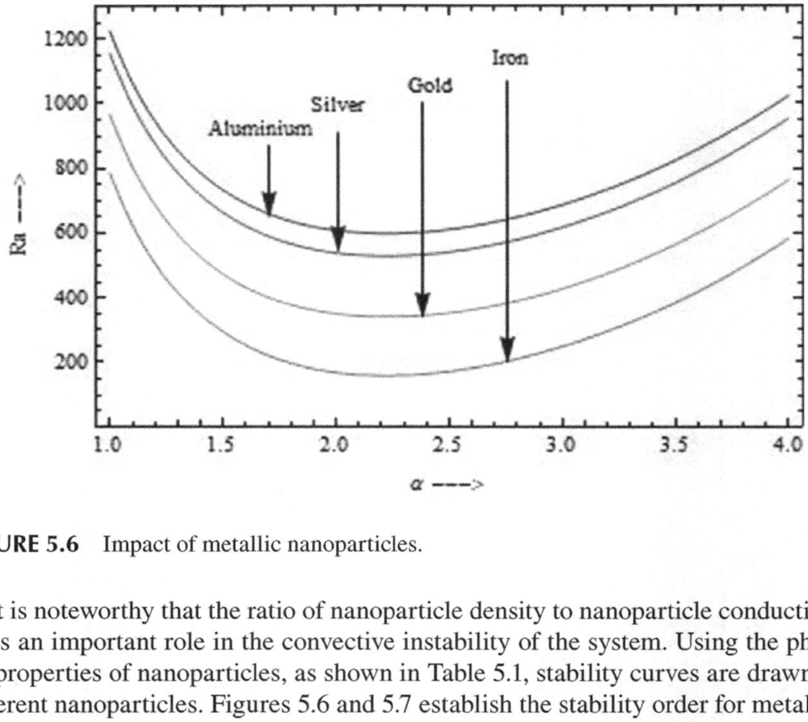


**FIGURE 5.6** Impact of metallic nanoparticles.

It is noteworthy that the ratio of nanoparticle density to nanoparticle conductivity plays an important role in the convective instability of the system. Using the physical properties of nanoparticles, as shown in Table 5.1, stability curves are drawn for different nanoparticles. Figures 5.6 and 5.7 establish the stability order for metals as aluminum > silver > gold > iron, while, for non-metals, the order is alumina > silicon oxide > copper oxide > titanium oxide. In addition, non-metallic nanofluids are less stable than metallic ones, which, in turn, are less stable than regular fluids.

## 5.7 CONCLUSIONS

This chapter considers the partial differential equations for instability in a nanofluid layer and these are solved for two different models. Model 1 considers the initial particle volume fraction to vary in the vertical direction, whereas a constant value of nanoparticle volume fraction at the initial state is considered under Model 2. The expressions for the thermal Rayleigh number are derived analytically and numerical computations are performed with the help of the software Mathematica.

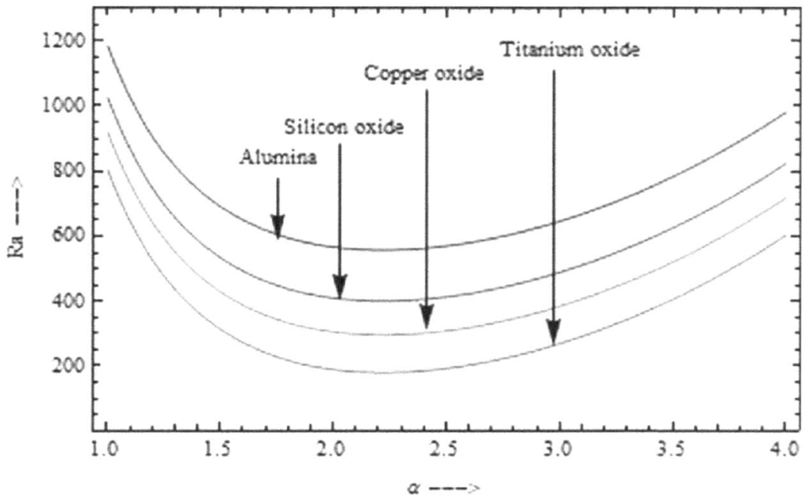


**FIGURE 5.7** Impact of non-metallic nanoparticles.

The instability of the system also depends on differences in nanoparticle volume fraction at the boundaries of the nanofluid layer, in addition to other parameters, if the initial nanoparticle volume fraction varies in the vertical direction, whereas this is not the case if nanoparticle volume fraction is considered to be constant at the basic state. Differences in the volume fractions at the boundaries make the system more unstable for top-heavy distributions than for a bottom-heavy situation under Model 1, whereas it has no impact on system stability for Model 2. Nanoparticle density destabilizes the system, whereas conductivity stabilizes it. The stability order for metals is established as aluminum > silver > gold > iron, whereas, for non-metals, it is alumina > silicon oxide > copper oxide > titanium oxide.

## REFERENCES


1. J.C. Maxwell, *Treatise on Electricity and Magnetism*, Clarendon Press, Oxford, 1873.
2. S. Choi, Enhancing thermal conductivity of fluids with nanoparticles: In D.A. Siginer, H.P. Wang (Eds.), *Development and Applications of Non-Newtonian Flows*. New York: American Society of Mechanical Engineers. FED- 231/MD-Vol. 66, 99–105, 1995. www.osti.gov/biblio/196525
3. S.K. Das and S.U.S. Choi, A review of heat transfer in nanofluids, *Advances in Heat Transfer*, 2009, 41, 81–197.
4. S.K. Das, S.U.S. Choi, W. Yu and T. Pradeep, *Nanofluids: Science and Technology*, Wiley, Hoboken, NY, 2008.
5. A. Zaman, N. Ali and I. Ali, Effect of nanoparticles (copper, silver) and slip on unsteady blood flow through a curved stenosed channel with aneurysm, *Thermal Science and Engineering Progress*, 2018, 5, 482–491. https://doi.org/10.1016/j.tsep.2018.02.004.
6. J. Buongiorno, Convective transport in nanofluids, *ASME Journal of Heat Transfer*, 2006, 128(3), 240–250.

7. S. Chandrasekhar, *Hydrodynamic and Hydromagnetic Stability*, Dover Publications, New York, 1981.
8. D.Y. Tzou, Instability of nanofluids in natural convection, *ASME Journal of Heat Transfer*, 2008, 130, 1–9.
9. D.Y. Tzou, Thermal instability of nanofluids in natural convection, *International Journal of Heat and Mass Transfer*, 2008, 51, 2967–2979.
10. D.A. Nield and A.V. Kuznetsov. The onset of convection in a horizontal nanofluid layer of finite depth. *European Journal of Mechanics: B/Fluids*, 2010, 29, 217–223.
11. U. Gupta, J. Ahuja and R.K. Wanchoo, Magneto convection in a nanofluid layer. *International Journal of Heat and Mass Transfer*, 2013, 64, 1163–1171.
12. S. Agarwal, B.S. Bhadauria and P.G. Siddheshwar, Thermal instability of a nanofluid saturating a rotating anisotropic porous medium, *Special Topics & Reviews in Porous Media: An International Journal*, 2011, 2(1), 53–64.
13. D.A. Nield and A.V. Kuznetsov, The onset of double-diffusive convection in a nanofluid layer, *International Journal of Heat and Fluid Flow*, 2011, 32(4), 771–776.
14. D. Yadav, G.S. Agrawal, and R. Bhargava, 2012, The onset of convection in a binary nanofluid saturated porous layer. *International Journal of Theoretical and Applied Multiscale Mechanics*, 2012, 2(3), 198–224.
15. U. Gupta, J. Sharma and V. Sharma, Instability of binary nanofluid with magnetic field, *Applied Mathematics and Mechanics*, 2015, 36(6), 693–706.
16. J. Sharma, U. Gupta and R.K. Wanchoo, Numerical study on binary nanofluid convection in a rotating porous layer, *Differential Equations and Dynamical Systems*, 2016, 25, 239–249. DOI 10.1007/s12591-015-0268-4.
17. R. Kumar, J. Sharma and Jyoti Sood, Rayleigh-Bénard cell formation of green synthesized nano-particles of silver and selenium, *Materials Today: Proceedings*, 2020, 28(3), 1781–1787.
18. D. Yadav, Impact of chemical reaction on the convective heat transport in nanofluid occupying in porous enclosures: A realistic approach, *International Journal of Mechanical Sciences*, 2019, 157–158, 357–373.
19. U. Gupta, J. Sharma and M. Devi, Casson nanofluid convection in an internally heated layer, *Materials Today: Proceedings*, 2020, 28(3), 1748–1752.
20. D.A. Nield and A.V. Kuznetsov, The onset of convection in a horizontal nanofluid layer of finite depth: A revised model, *International Journal of Heat and Mass Transfer*, 2014, 77, 915–918.
21. R. Chand and G.C. Rana, Thermal instability in a Brinkman porous medium saturated by nanofluid with no nanoparticle fluxon boundaries. *Special Topics & Reviews in Porous Media: An International Journal*, 2014, 5, 277–86.
22. S. Agarwal, Natural convection in a nanofluid-saturated rotating porous layer: A more realistic approach, *Transport in Porous Media*, 2014, 104, 581–92.
23. R. Chand, G.C. Rana. Magneto convection in a layer of nanofluid in porous medium: A more realistic approach. *Journal of Nanofluids*, 2015, 4, 196–202.
24. J. Sharma, U. Gupta and V. Sharma, Modified model for binary nanofluid convection with constant nanoparticle volume fraction, *Journal of Applied Fluid Mechanics*, 2017, 10(5), 1387–1395.
25. J. Ahuja and J. Sharma, Rayleigh–Bénard instability in nanofluids: A comprehensive review, *Micro and Nano Systems Letters*, 2020, 8, 21 https://doi.org/10.1186/s40486-020-00123-y.

# 6 Analysis of Prey–Predator Model

*Purnima Pandit, Payal Singh, and Tanvi Patel*

## CONTENTS



## 6.1 INTRODUCTION

The prey–predator model that we consider here contains a system of nonlinear differential equations and is used for describing the behavior of biological systems. In this classical model, two species are considered: one species (the prey) depends on plant material available in the forest and elsewhere, and the other species (the predator) eats the prey species. When food is available in abundance, then the population of the prey species increases; when they come in contact with predators and/or suffer natural death, their population size decreases. Similarly, when predators get their food (prey items), their population size increases, and when there is a shortage of prey items and/or natural death occurs, their population size decreases.

The mathematical model is given as follows,

$$\begin{cases} \dot{x} = ax - bxy \\ \dot{y} = -cy + dxy \end{cases} \tag{6.1}$$

with initial conditions $x_0$ and $y_0$,

where $a$ represents the growth rate of the prey species ($x$), $b$ represents the decay rate of the prey population when they interact with predators, $c$ represents the decay rate of the predator population ($y$) and $d$ represents the growth rate of the predators when they get enough food (prey).

Many authors [1–8] have solved Eq. (6.1) by different techniques and converted this system into modified forms by applying different scenarios to real-life problems.

Due to the competitive behavior of Eq. (6.1), the application of such a model is not limited to systems of only biological importance. It is also used in finance, engineering, epidemic modeling, and many more areas.

This chapter focuses on the solution techniques for Eq. (6.1). We see that Eq. (6.1) contains a nonlinear term; at times, it is not possible to obtain an exact solution to nonlinear differential equations. In this chapter, Eq. (6.1) is solved by taking three cases to observe the interaction between prey and predators. The first case reflects that there is no interaction between prey and predator: mathematically, $b = d = 0$. In the second case, this nonlinear term which describes the interaction between prey and predator in Eq. (6.1) is linearized about the equilibrium point and we obtain a closed-form solution. In the third and last case, Eq. (6.1) is solved by a numerical technique.

There are six sections to this chapter. Section 6.1 is the Introduction, Section 6.2 contains descriptions of the techniques used, and Section 6.3 contains a stability analysis for a general nonlinear dynamical system, whereas Section 6.4 describes stability analysis for the prey–predator system. Section 6.5 deals with applications of the prey–predator model, followed by Section 6.6 (Results and Discussion) and Section 6.7 (Conclusion).

## 6.2 DESCRIPTION OF METHOD

Consider Eq. (6.1) in matrix form,

$$\begin{bmatrix} \dot{x} \\ \dot{y} \end{bmatrix} = \begin{bmatrix} a & 0 \\ 0 & -c \end{bmatrix} \begin{bmatrix} x \\ y \end{bmatrix} + \begin{pmatrix} -bxy \\ dxy \end{pmatrix}$$

with initial condition $\begin{bmatrix} x_0 & y_0 \end{bmatrix}^T$. In the following, we proceed to obtain the solution of this equation by taking various cases.

### 6.2.1 Case 1

When we ignore the nonlinear term, we have the following equation,

$$\begin{bmatrix} \dot{x} \\ \dot{y} \end{bmatrix} = \begin{bmatrix} a & 0 \\ 0 & -c \end{bmatrix} \begin{bmatrix} x \\ y \end{bmatrix}$$

with initial condition $\begin{bmatrix} x_0 & y_0 \end{bmatrix}^T$.

We can write the above equation in the following manner:

$$\dot{X}(t) = AX(t) \tag{6.2}$$

where $A$ is a $2 \times 2$ matrix and $X$ is a $2 \times 1$ column vector. The existence result of solution for Eq. (6.2) is given in the following theorem.

### 6.2.2 Theorem 1

If matrix $A$ is continuous, then the solution of Eq. (6.2) is given as follows:

$$\begin{bmatrix} x \\ y \end{bmatrix} = \begin{bmatrix} x_0 e^{at} \\ y_0 e^{-ct} \end{bmatrix}$$

**Proof:**

Since matrix $A$ contains only constant terms, it is continuous. Now, to solve an equation such as Eq. (6.2), we need to find eigenvalues and eigenvectors of the equation, i.e., $\lambda_1 = a, \lambda_2 = -c$ and $v_1 = \begin{bmatrix} 1 \\ 0 \end{bmatrix}$ and $v_2 = \begin{bmatrix} 0 \\ 1 \end{bmatrix}$ are the eigenvectors corresponding to $\lambda_1 = a$ and $\lambda_2 = -c$, respectively.

So, the solution of Eq. (6.2) is given as:

$$\begin{bmatrix} x \\ y \end{bmatrix} = c_1 v_1 e^{\lambda_1 t} + c_2 v_2 e^{\lambda_2 t}$$

where $c_1$ and $c_2$ are arbitrary constants. Values of these constants are obtained using the initial conditions.

$$\begin{bmatrix} x \\ y \end{bmatrix} = x_0 \begin{bmatrix} 1 \\ 0 \end{bmatrix} e^{at} + y_0 \begin{bmatrix} 0 \\ 1 \end{bmatrix} e^{-ct}$$

$$\begin{bmatrix} x \\ y \end{bmatrix} = \begin{bmatrix} x_0 e^{at} \\ y_0 e^{-ct} \end{bmatrix}$$

### 6.2.3 Case 2

In the second case, we linearize the nonlinear term about the equilibrium point. To find the equilibrium point, we put $\dot{x}=0$ and $\dot{y}=0$ in Eq. (6.1). It gives two equilibrium points; one is (0,0) and the other is $\left(\frac{c}{d},\frac{a}{b}\right)$. If we linearize Eq. (6.1) about point (0,0), then the nonlinear part will disappear, so we choose another point $\left(\frac{c}{d},\frac{a}{b}\right)$. By using Taylor's expansion about the equilibrium point $\left(\frac{c}{d},\frac{a}{b}\right)$ and ignoring the higher terms, Eq. (6.1) is converted into the following equation:

$$\begin{bmatrix}\dot{x}\\ \dot{y}\end{bmatrix}=\begin{bmatrix}0 & \frac{-bc}{d}\\ \frac{ad}{b} & 0\end{bmatrix}\begin{bmatrix}x\\ y\end{bmatrix}+\begin{bmatrix}\frac{ac}{d}\\ \frac{-ac}{b}\end{bmatrix}$$

with initial condition $\begin{bmatrix}x_0 & y_0\end{bmatrix}^T$.

Or,

$$\dot{X}(t)=PX(t)+Q \tag{6.3}$$

where

$$P=\begin{bmatrix}0 & \frac{-bc}{d}\\ \frac{ad}{b} & 0\end{bmatrix},\ Q=\begin{bmatrix}\frac{ac}{d}\\ \frac{-ac}{b}\end{bmatrix}\text{ and }X=\begin{bmatrix}x\\ y\end{bmatrix}$$

The solution of Eq. (6.3) is given in the next theorem (Section 6.2.4).

### 6.2.4 Theorem 2

If $X(t)$, $P$, and $Q$ are all continuous functions, then the solution of Eq. (6.3) is given as follows:

$$y(t)=\frac{1}{ac}+\left(\frac{\sqrt{ad}\,x_0}{\sqrt{cb}}-\frac{\sqrt{ac}}{b}\right)\sin\left(\sqrt{ac}\,t\right)+\left(\frac{y_0}{\sqrt{cb}}-\frac{1}{(ac)^{\frac{3}{2}}}\right)\cos\left(\sqrt{ac}\,t\right)$$

$$x(t)=-\frac{c}{d}+\left(\frac{\sqrt{ac}}{d}+\frac{\sqrt{c}}{\sqrt{a}}\frac{b}{d}y_0\right)\sin\left(\sqrt{ac}\,t\right)+\left(x_0+\frac{c}{d}\right)\cos\left(\sqrt{ac}\,t\right)$$

**Proof:**

Because of the constant terms in $P$ and $Q$, both are continuous matrices, so we solve Eq. (6.3) by the Laplace transform method.

Applying the Laplace transform method to both sides of Eq. (6.3),

$$L(\dot{x}) = L\left(\frac{-bc}{d}y\right) + L\left(\frac{ac}{d}\right)$$

$$L(\dot{y}) = L\left(\frac{ad}{b}x\right) + L\left(\frac{-ac}{b}\right)$$

Here, we use the linear property of the Laplace transform and some equations,

$$L(\dot{x}) = sx(s) - x_0,\ L(\dot{y}) = sx(y) - y_0,\ L(1) = \frac{1}{S}$$

$$sx(s) - x_0 = \frac{-bc}{d}y(s) + \frac{ac}{ds}$$

$$sx(s) + \frac{bc}{d}y(s) = x_0 + \frac{ac}{ds} \tag{6.4}$$

$$sy(s) - \frac{ad}{b}x(s) = y_0 - \frac{ac}{bs} \tag{6.5}$$

Eq. (6.4) is multiplied by $\frac{ad}{b}$, and Eq. (6.5) is multiplied by $s$, and we add both new equations (6.4) and (6.5) and take the inverse Laplace transform, to obtain,

$$y(t) = \frac{1}{ac} + \left(\frac{\sqrt{ad}\,x_0}{\sqrt{cb}} - \frac{\sqrt{ac}}{b}\right)\sin\left(\sqrt{ac}\,t\right) + \left(\frac{y_0}{\sqrt{cb}} - \frac{1}{(ac)^{\frac{3}{2}}}\right)\cos\left(\sqrt{ac}\,t\right)$$

Now, Eq. (6.5) is multiplied by $\frac{bc}{d}$, and Eq. (6.4) is multiplied by $s$, and we subtract them and take the inverse Laplace transform, to obtain

$$x(t) = -\frac{c}{d} + \left(\frac{\sqrt{ac}}{d} + \frac{\sqrt{c}}{\sqrt{a}}\frac{b}{d}y_0\right)\sin\left(\sqrt{ac}\,t\right) + \left(x_0 + \frac{c}{d}\right)\cos\left(\sqrt{ac}\,t\right)$$

### 6.2.5 Case 3

In case 3, we consider that the nonlinear term is the Lipschitz continuous function and apply Runge-Kutta fourth-order derivative (numerical technique).

This system contains coupled nonlinear differential equations so, for the numerical scheme, in each iteration, we use simultaneous slope values. The solving technique is given as follows:

Let $f(t,x,y) = \dot{x}(t) = ax - bxy$, $g(t,x,y) = \dot{y}(t) = -cy + dxy$ in Eq. (6.1).

We obtain $k_1$ and $l_1$ by using values from Eq. (6.1),

$$k_1 = hf\left(t_n, x_n, y_n\right)$$

$$k_1 = h\left(ax_n - bx_ny_n\right)$$

$$l_1 = g\left(t_n, x_n, y_n\right)$$

$$l_1 = h\left(-cy_n + dx_ny_n\right)$$

By using values of $k_1$, $l_1$ from Eq. (6.1), we can determine $k_2$ and $l_2$.

$$k_2 = f\left(\left(t_n + 0.5h\right), \left(x_n + 0.5k_1\right), \left(y_n + 0.5l_1\right)\right)$$

$$k_2 = a\left(x_n + 0.5k_1\right) - b\left(x_n + 0.5k_1\right)\left(y_n + 0.5l_1\right)$$

$$k_2 = a\left(x_n + 0.5h\left(ax_n - bx_ny_n\right)\right) - b\left(x_n + 0.5h\left(ax_n - bx_ny_n\right)\right)\left(y_n + 0.5h\left(-cy_n + dx_ny_n\right)\right)$$

$$l_2 = g\left(\left(t_n + 0.5h\right), \left(x_n + 0.5k_1\right), \left(y_n + 0.5l_1\right)\right)$$

$$l_2 = h\left(-c(y_n + 0.5h\left(-cy_n + dx_ny_n\right)\right) + d\left(x_n + 0.5h\left(ax_n - bx_ny_n\right)(y_n + 0.5h\left(-cy_n + dx_ny_n\right)\right)$$

By using values of $k_2$ and $l_2$ from Eq. (6.1), we can determine $k_3$ and $l_3$.

$$k_3 = f\left(\left(t_n + 0.5h\right), \left(x_n + 0.5k_2\right), \left(y_n + 0.5l_2\right)\right)$$

$$\begin{aligned} k_3 = {} & a\left(x_n + 0.5a\left(x_n + 0.5h\left(ax_n - bx_ny_n\right)\right) - b\left(x_n + 0.5h\left(ax_n - bx_ny_n\right)\right)(y_n + 0.5h\left(-cy_n + dx_ny_n\right)\right) \\ & - b\left(x_n + 0.5a\left(x_n + 0.5h\left(ax_n - bx_ny_n\right)\right) - b\left(x_n + 0.5h\left(ax_n - bx_ny_n\right)\right)(y_n + 0.5h\left(-cy_n + dx_ny_n\right)\right) \\ & (y_n + 0.5h\left(-c(y_n + 0.5h\left(-cy_n + dx_ny_n\right)\right) + d\left(x_n + 0.5h\left(ax_n - bx_ny_n\right)(y_n + 0.5h\left(-cy_n + dx_ny_n\right)\right) \end{aligned}$$

$$l_3 = g\left(\left(t_n + 0.5h\right), \left(x_n + 0.5k_2\right), \left(y_n + 0.5l_2\right)\right)$$

$$\begin{aligned} l_3 = {} & -c(y_n + 0.5h\left(-c(y_n + 0.5h\left(-cy_n + dx_ny_n\right)\right) + d(x_n + 0.5h\left(ax_n - bx_ny_n\right) \\ & (y_n + 0.5h\left(-cy_n + dx_ny_n\right) \\ & + d(\left(x_n + 0.5a\left(x_n + 0.5h\left(ax_n - bx_ny_n\right)\right) - b\left(x_n + 0.5h\left(ax_n - bx_ny_n\right)\right)(y_n + 0.5h\left(-cy_n + dx_ny_n\right)\right) \\ & (y_n + 0.5h\left(-c(y_n + 0.5h\left(-cy_n + dx_ny_n\right)\right) \\ & + d\,(x_n + 0.5h\left(ax_n - bx_ny_n\right)(y_n + 0.5h - \left(-cy_n + dx_ny_n\right) \end{aligned}$$

By using values of $k_3$, $l_3$ from Eq. (6.1), we can determine $k_4$ and $l_4$

$$k_4 = f\left(\left(t_n + h\right), \left(x_n + k_3\right), \left(y_n + l_3\right)\right)$$

$$k_4 = a(x_n + a\left(x_n + 0.5a\left(x_n + 0.5h\left(ax_n - bx_n y_n\right)\right) - b\left(x_n + 0.5h\left(ax_n - bx_n y_n\right)\right)\right.$$

$$(y_n + 0.5h\left(-cy_n + dx_n y_n\right)\Big)$$

$$-b\left(x_n + 0.5a\left(x_n + 0.5h\left(ax_n - bx_n y_n\right)\right) - b\left(x_n + 0.5h\left(ax_n - bx_n y_n\right)\right)(y_n + 0.5h\left(-cy_n + dx_n y_n\right)\right)$$

$$(y_n + 0.5h\left(-c(y_n + 0.5h\left(-cy_n + dx_n y_n\right)\right) + d\left(x_n + 0.5h\left(ax_n - bx_n y_n\right)(y_n + 0.5h\left(-cy_n + dx_n y_n\right)\right)$$

$$-b(x_n + a\left(x_n + 0.5a\left(x_n + 0.5h\left(ax_n - bx_n y_n\right)\right) - b\left(x_n + 0.5h\left(ax_n - bx_n y_n\right)\right)(y_n + 0.5h\left(-cy_n + dx_n y_n\right)\right)$$

$$-b\left(x_n + 0.5a\left(x_n + 0.5h\left(ax_n - bx_n y_n\right)\right) - b\left(x_n + 0.5h\left(ax_n - bx_n y_n\right)\right)(y_n + 0.5h\left(-cy_n + dx_n y_n\right)\right)$$

$$(y_n + 0.5h\left(-c(y_n + 0.5h\left(-cy_n + dx_n y_n\right)\right)$$

$$+d\left(x_n + 0.5h\left(ax_n - bx_n y_n\right)(y_n + 0.5h\left(-cy_n + dx_n y_n\right)\right)(y_n + -c(y_n + 0.5h\left(-c(y_n + 0.5h\left(-cy_n + dx_n y_n\right)\right)$$

$$+d\left(x_n + 0.5h\left(ax_n - bx_n y_n\right)(y_n + 0.5h\left(-cy_n + dx_n y_n\right)\right)$$

$$+d\left(\left(x_n + 0.5a\left(x_n + 0.5h\left(ax_n - bx_n y_n\right)\right) - b\left(x_n + 0.5h\left(ax_n - bx_n y_n\right)\right)(y_n + 0.5h\left(-cy_n + dx_n y_n\right)\right)\right.$$

$$(y_n + 0.5h\left(-c(y_n + 0.5h\left(-cy_n + dx_n y_n\right)\right)\Big)$$

$$+d\left(x_n + 0.5h\left(ax_n - bx_n y_n\right)(y_n + 0.5h\left(-cy_n + dx_n y_n\right)\right)$$

$$l_4 = g\left(\left(t_n + h\right), \left(x_n + k_3\right), \left(y_n + l_3\right)\right)$$

$$l_4 = -c(y_n + -c(y_n + 0.5h\left(-c(y_n + 0.5h\left(-cy_n + dx_n y_n\right)\right)$$

$$+d\left(x_n + 0.5h\left(ax_n - bx_n y_n\right)(y_n + 0.5h\left(-cy_n + dx_n y_n\right)\right)$$

$$+d\left(\left(x_n + 0.5a\left(x_n + 0.5h\left(ax_n - bx_n y_n\right)\right) - b\left(x_n + 0.5h\left(ax_n - bx_n y_n\right)\right)(y_n + 0.5h\left(-cy_n + dx_n y_n\right)\right)\right.$$

$$\left(y_n + 0.5h\left(-c(y_n + 0.5h\left(-cy_n + dx_n y_n\right)\right)\right.$$

$$+d\left(x_n + 0.5h\left(ax_n - bx_n y_n\right)(y_n + 0.5h\left(-cy_n + dx_n y_n\right)\right)\Big)$$

$$+d(x_n + a\left(x_n + 0.5a\left(x_n + 0.5h\left(ax_n - bx_n y_n\right)\right)\right.$$

$$-b\left(x_n + 0.5h\left(ax_n - bx_n y_n\right)\right)(y_n + 0.5h\left(-cy_n + dx_n y_n\right)\Big)$$

$$-b\left(x_n + 0.5a\left(x_n + 0.5h\left(ax_n - bx_n y_n\right)\right) - b\left(x_n + 0.5h\left(ax_n - bx_n y_n\right)\right)(y_n + 0.5h\left(-cy_n + dx_n y_n\right)\right)$$

$$(y_n + 0.5h\left(-c(y_n + 0.5h\left(-cy_n + dx_n y_n\right)\right) + d\left(x_n + 0.5h\left(ax_n - bx_n y_n\right)(y_n + 0.5h\left(-cy_n + dx_n y_n\right)\right)$$

$$(y_n + -c(y_n + 0.5h(-c(y_n + 0.5h(-cy_n + dx_n y_n)) + d(x_n + 0.5h(ax_n - bx_n y_n)(y_n + 0.5h(-cy_n + dx_n y_n))$$

$$+d((x_n + 0.5a(x_n + 0.5h(ax_n - bx_n y_n)) - b(x_n + 0.5h(ax_n - bx_n y_n))(y_n + 0.5h(-cy_n + dx_n y_n))$$

$$(y_n + 0.5h(-c(y_n + 0.5h(-cy_n + dx_n y_n)) + d(x_n + 0.5h(ax_n - bx_n y_n)(y_n + 0.5h(-cy_n + dx_n y_n))$$

From the above solution technique, we can find a numerical solution to Eq. (6.1), with desired values of $n = 1, 2, 3\ldots$

$$x_{n+1} = x_n + \frac{(k_1 + 2k_2 + 2k_3 + k_4)}{6}$$

$$y_{n+1} = y_n + \frac{(l_1 + 2l_2 + 2l_3 + l_4)}{6}$$

## 6.3 STABILITY ANALYSIS

For a general nonlinear dynamical system,

$$\begin{cases} \dot{x}(t) = f(t, x, y) \\ \dot{y}(t) = g(t, x, y) \end{cases} \tag{6.6}$$

Let $f(t, x, y)$ and $g(t, x, y)$ both be smoothed functions in the equation and $(x_e, y_e)$ be an equilibrium point of the above dynamical system. If perturbation is performed in $p(t)$ and $q(t)$ in $(x_e, y_e)$, we have $x_e + p(t)$ and $y_e + q(t)$. These points are also solutions to the above dynamical system, as they satisfy the above equation,

$$\frac{d(x_e + p(t))}{dt} = f(t, x_e + p(t), y_e + q(t))$$

$$\frac{d(y_e + q(t))}{dt} = g(t, x_e + p(t), y_e + q(t))$$

### 6.3.1 Theorem 3

Eq. (6.6) has a stable solution, if perturbations $p(t)$, $q(t)$ tend to zero as $t \to \infty$

**Proof:**

Since $f(t, x, y)$ and $g(t, x, y)$ are both smoothed functions in Eq. (6.6), we can expand these functions by Taylor's theorem,

$$\frac{d(p(t))}{dt} = f(x_e, y_e) + \frac{\partial f}{\partial x}_{(x_e, y_e)} p(t) + \frac{\partial f}{\partial y}_{(x_e, y_e)} q(t)$$

$$\frac{d(q(t))}{dt} = g(x_e, y_e) + \frac{\partial g}{\partial x}_{(x_e, y_e)} p(t) + \frac{\partial g}{\partial y}_{(x_e, y_e)} q(t)$$

Now, $f(x_e, y_e) = g(x_e, y_e) = 0$, so we have,

$$\begin{cases} \dfrac{d(p(t))}{dt} = \dfrac{\partial f}{\partial x} p(t) + \dfrac{\partial f}{\partial y} q(t) \\ \dfrac{d(q(t))}{dt} = \dfrac{\partial g}{\partial x} p(t) + \dfrac{\partial g}{\partial y} q(t) \end{cases} \tag{6.7}$$

Eq. (6.6) becomes a homogeneous system, Eq. (6.7), by making appropriate assumptions. Jacobian of Eq. (6.7),

$$J = \begin{bmatrix} \dfrac{\partial f}{\partial x} & \dfrac{\partial f}{\partial y} \\ \dfrac{\partial g}{\partial x} & \dfrac{\partial g}{\partial y} \end{bmatrix}$$

For convenience, let's take $\dfrac{\partial f}{\partial x} = \alpha, \dfrac{\partial f}{\partial y} = \beta, \dfrac{\partial g}{\partial x} = \gamma, \dfrac{\partial g}{\partial y} = \delta,$

$$J = \begin{bmatrix} \alpha & \beta \\ \gamma & \delta \end{bmatrix}$$

Eigenvalues of this matrix can be found from the following equation,

$$\lambda^2 - (\alpha + \delta)\lambda + \alpha\delta - \beta\gamma = 0$$

Let $u = (\alpha + \delta)$ and $= \alpha\delta - \beta\gamma$,

$$\lambda^2 - u\lambda + v = 0$$

Now, $\lambda_1, \lambda_2 = \dfrac{u \pm \sqrt{u^2 - 4v}}{2}$ and the corresponding eigenvectors are $v_1, v_2$.

So, the solution of Eq. (6.7) is,

$$\begin{bmatrix} p \\ q \end{bmatrix} = c_1 v_1 e^{\lambda_1} t + c_1 v_1 e^{\lambda_1} t \tag{6.8}$$

The following conditions exist for the stability of the equation:

- If $u$ is negative and $\sqrt{u^2 - 4v} > 0$ then both eigenvalues will be negative, hence the equation will be stable, because the solution of Eq. (6.8) will contain $\exp(-\lambda t)$ so, as $t \to \infty$, the perturbation becomes zero, which is the required condition for stability.
- If both eigenvalues are positive, then Eq. (6.6) becomes unstable.
- If $u^2 - 4v < 0$, then we have imaginary eigenvalues, which give three cases: real part positive (solution is unstable and undamped), real part negative (solution is stable and damped), and real part zero (system behaves as an undamped oscillator).

## 6.4 STABILITY ANALYSIS FOR PREY–PREDATOR MODEL

From theorem (3) (Section 6.3.1), we can find the Jacobian of Eq. (6.1). For Eq. (6.1), we have two equilibrium points, (0,0) and $\left(\frac{c}{d}, \frac{a}{b}\right)$.

For (0,0), eigenvalues are $a$ and $-c$, and the corresponding eigenvectors are $\begin{bmatrix} k \\ 0 \end{bmatrix}$ and $\begin{bmatrix} 0 \\ k \end{bmatrix}$, respectively, where $k$ is an arbitrary constant. So, the point (0,0) is the saddle point.

For $\left(\frac{c}{d}, \frac{a}{b}\right)$, eigenvalues are $\pm i\sqrt{ac}$, then, from Eq. (6.3), Eq. (1) behaves like an undamped oscillator. The solution of Eq. (6.1) contains a periodic function, which gives the cyclic behavior of Eq.(6.1).

## 6.5 APPLICATIONS

### 6.5.1 Disease Model

This application is taken from [9].

$$\begin{cases} \dot{x} = -kxy \\ \dot{y} = kxy - cy \\ z = N - (x + y) \end{cases} \tag{6.9}$$

with initial condition $x_0$, $y_0$ and $z_0 = 0$, $N = x_0 + y_0$

$$k = 0.005, c = 0.9, x_0 = 950, y_0 = 50$$

where $x(t)$ = number of people who are uninfected but may become infected,
$y(t)$ = number of people who are infected and may spread the disease, and
$z(t)$ = people who are dead, immune, or cannot spread the disease.

So, $N = x(t) + y(t) + z(t)$

**Solution:**

This Eq. (6.9) has been solved by all three cases, given in Section 6.2.

#### 6.5.1.1 Case 1

When the nonlinear term is ignored, then

$$\dot{x} = 0, \dot{y} = -cy; k = 0.005, c = 0.9, x_0 = 950, y_0 = 50$$

Solution of above equations:

$$x(t) = 950, y(t) = 50e^{-0.9t}$$

The dynamics of case 1 is given in Figure 6.1 and Table 6.1.

#### 6.5.1.2 Case 2

When the nonlinear term is linearized about the equilibrium point, Eq. (6.9) becomes converted into $\dot{x} = -0.9y, \dot{y} = 0; k = 0.005, c = 0.9, x_0 = 950, y_0 = 50$

The dynamics of case 2 is given in Figure 6.2 and Table 6.2.

#### 6.5.1.3 Case 3

When all terms are considered in Eq. (6.9), the dynamics of case 3 is given in Figure 6.3 and Table 6.3.

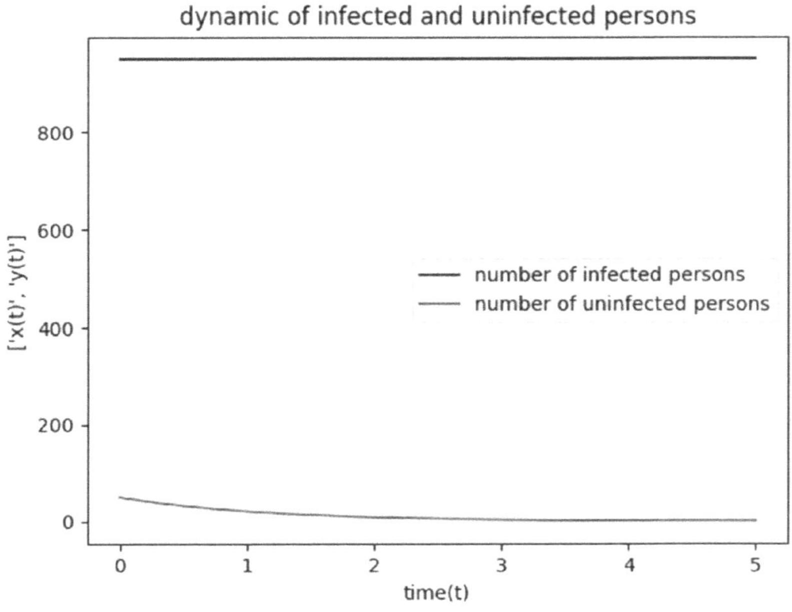


**FIGURE 6.1** When the nonlinear term is neglected.

**TABLE 6.1**
**When the Nonlinear Term is Ignored**

| Time ($t$) | Number of uninfected people ($x$) | Number of infected people ($y$) |
|---|---|---|
| 0 | 950 | 50 |
| 1 | 950 | 20.3285 |
| 2 | 950 | 8.2649 |
| 3 | 950 | 3.3603 |
| 4 | 950 | 1.3662 |
| 5 | 950 | 0.5554 |

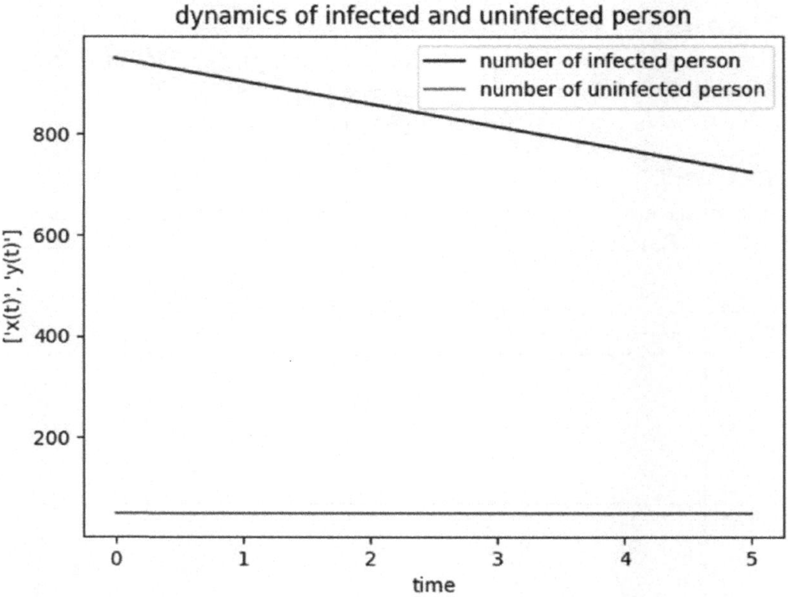


**FIGURE 6.2** When the nonlinear term is linearized.

### 6.5.2 Numerical Illustration

Consider the following example of the prey–predator model:

$$\begin{cases} \dot{x} = 0.1x - 0.005xy \\ \dot{y} = -0.4y + 0.008xy \end{cases} \tag{6.10}$$

$$x_0 = 130, y_0 = 40$$

**TABLE 6.2**
**When the Nonlinear Term is Linearized**

| Time ($t$) | Number of uninfected people ($x$) | Number of infected people ($y$) |
|---|---|---|
| 0 | 950 | 50 |
| 1 | 905 | 50 |
| 2 | 860 | 50 |
| 3 | 815 | 50 |
| 4 | 770 | 50 |
| 5 | 725 | 50 |

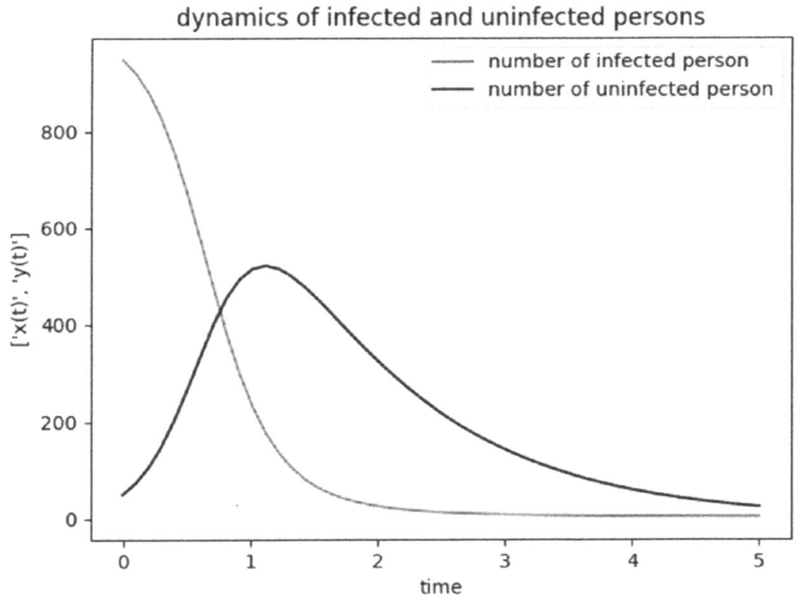


**FIGURE 6.3** When the numerical scheme is applied.

### 6.5.2.1 Case 1

When we ignore the nonlinear term, Eq. (6.8) becomes converted into the following equation:

$$\begin{cases} \dot{x} = 0.1x \\ \dot{y} = -0.4y \end{cases} \tag{6.11}$$

with initial condition $x_0 = 130, y_0 = 4$

From Section 6.2, the solution of Eq. (6.10) is given in Figure 6.4 and Table 6.4.

**TABLE 6.3**
**When the Numerical Scheme is Applied**

| Time ($t$) | Number of uninfected people ($x$) | Number of infected people ($y$) |
|---|---|---|
| 0 | 950 | 50 |
| 0.0130487606831667 | 946.827171718900 | 52.5706543694719 |
| 0.0260975213663333 | 943.502996146759 | 55.2617637618087 |
| 0.0391462820495000 | 940.021627029553 | 58.0777347878430 |
| 0.0521950427326666 | 936.377127214582 | 61.0230112456377 |
| 0.113015367857654 | 917.069944787758 | 76.5785999248012 |
| 0.173835692982642 | 893.459285772626 | 95.4951806522929 |
| 0.234656018107630 | 864.972923317156 | 118.151667344044 |
| 0.295476343232618 | 831.160023868247 | 144.787605907546 |
| 0.376281942634740 | 777.564619979669 | 186.393700325882 |
| 0.457087542036862 | 714.334577433357 | 234.355618169416 |
| 0.537893141438984 | 642.973193985640 | 286.776129156155 |
| 0.618698740841106 | 566.361441681294 | 340.574215345090 |
| 0.716203374832596 | 472.477162654015 | 401.787144125838 |
| 0.813708008824086 | 383.414741356468 | 453.235212406172 |
| 0.911212642815577 | 304.279484638039 | 490.875922243329 |
| 1.00871727680707 | 237.952622439857 | 513.055766871196 |
| 1.10558538044434 | 185.239743942688 | 520.647681483385 |
| 1.20245348408161 | 143.962319250108 | 516.657131918483 |
| 1.29932158771887 | 112.175422930201 | 503.876884714817 |

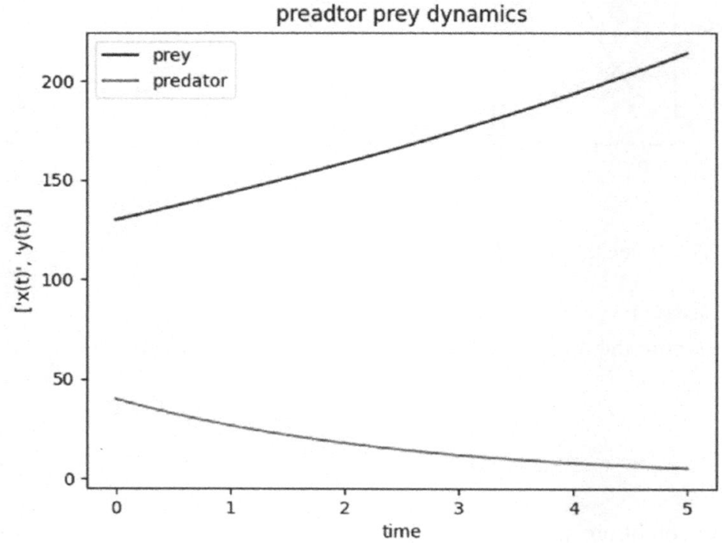


**FIGURE 6.4** When the nonlinear term is ignored.

**TABLE 6.4**
**When Nonlinear Term is Ignored**

| Time $(t)$ | $x$ No. of prey | $y$ No. of predators |
|---|---|---|
| 0 | 130 | 40 |
| 0.1 | 131.33332169 | 38.40021812 |
| 0.4 | 135.41589911 | 33.97463305 |
| 0.8 | 141.05742872 | 28.85689205 |
| 1.2 | 146.93398877 | 24.51005789 |
| 1.6 | 153.05537078 | 20.81800553 |
| 2.0 | 159.43177423 | 17.68210213 |
| 2.4 | 166.07382352 | 15.01857315 |
| 2.8 | 172.99258564 | 12.75626264 |
| 3.2 | 180.19958867 | 10.83473342 |
| 3.6 | 187.70684094 | 9.20265218 |
| 4.0 | 195.52685107 | 7.81641811 |
| 4.4 | 203.67264878 | 6.6389983 |
| 4.8 | 212.15780663 | 5.63893818 |
| 5 | 214.33376516 | 5.41341132 |

#### 6.5.2.2 Case 2

In this case, we linearize the nonlinear term around the equilibrium point. For Eq. (6.10), the equilibrium point is (50,20). From Section (6.2), we have

$$x(t) = 50 + 80\cos(0.2t) - 25\sin(0.2t)$$

$$y(t) = 20 + 20\cos(0.2t) + 64\sin(0.2t)$$

Figure 6.5 and Table 6.5 show the result based on the above solution.

#### 6.5.2.3 Case 3

In this case, we obtain the solution by a numerical technique, and the dynamics of case 3 is given in Figure 6.6 and Table 6.6.

## 6.6 RESULTS AND DISCUSSION

We have solved one application and one numerical illustration based on the prey–predator model. The application based on infected and uninfected people is given by Eq. (6.9). In case 1, when we did not consider the nonlinear term, then we can see, from Figure 6.1 and Table 6.1, that the population of uninfected people becomes constant and the population of infected people decreases exponentially. In case 2, when the equation is linearized about the equilibrium point, then the dynamics of infected people and uninfected people are shown in Figure 6.2 and Table 6.2. In case 3, when

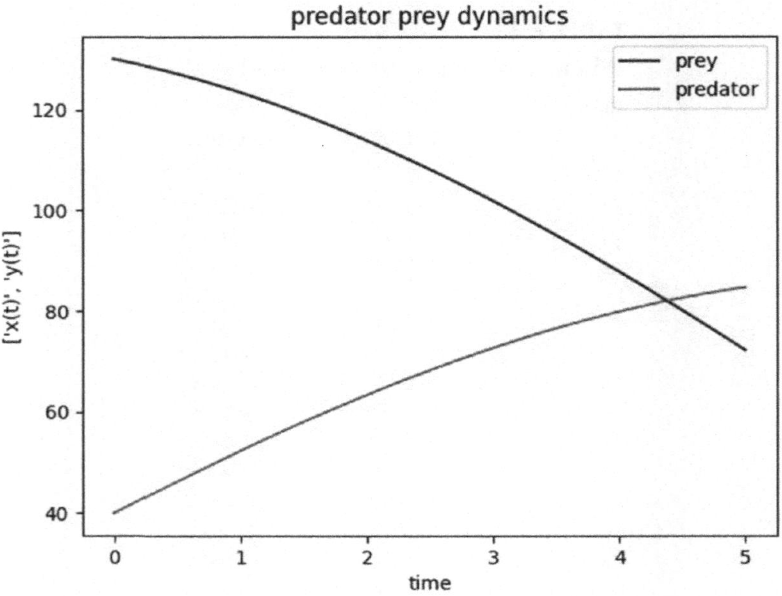


**FIGURE 6.5** When the nonlinear term is linearized.

**TABLE 6.5**
**When the Nonlinear Term is Linearized**

| Time (*t*) | *x* No. of prey | *y* No. of predators |
|---|---|---|
| 0 | 130 | 40 |
| 0.1 | 129.47317244 | 41.3018664 |
| 0.4 | 127.69504222 | 45.1520867 |
| 0.8 | 124.87262044 | 50.13665626 |
| 1.2 | 121.55153275 | 54.92050996 |
| 1.6 | 117.75389824 | 59.47178643 |
| 2.0 | 113.50500943 | 63.76017302 |
| 2.4 | 108.8331655 | 67.7571086 |
| 2.8 | 103.76948211 | 71.43597298 |
| 3.2 | 98.34768379 | 74.77226406 |
| 3.6 | 92.60388039 | 77.74376133 |
| 4.0 | 86.576327 | 80.33067417 |
| 4.4 | 80.3051683 | 82.51577324 |
| 4.8 | 73.83217147 | 84.28450537 |
| 5 | 72.18741168 | 84.66018847 |

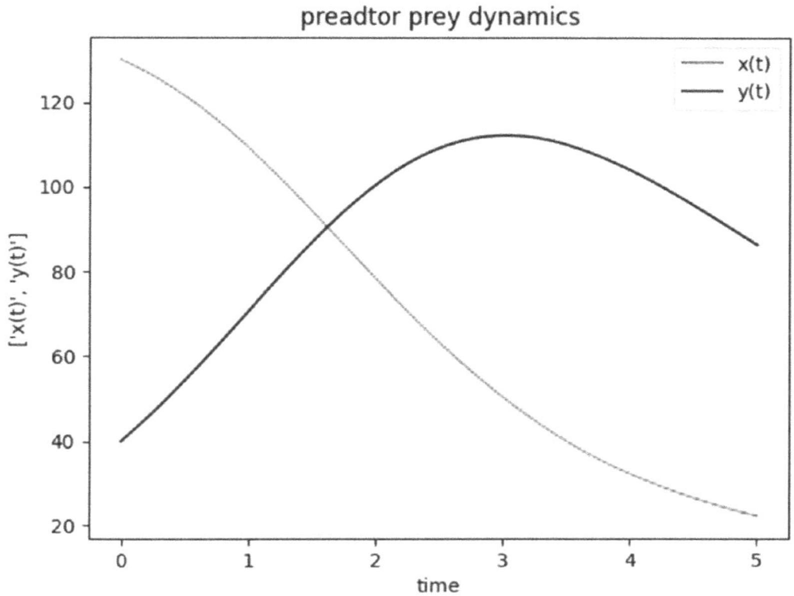


**FIGURE 6.6** When the numerical scheme is applied.

**TABLE 6.6**
**When the Numerical Scheme is Applied**

| Time ($t$) | $x$ No. of prey | $y$ No. of predators |
|---|---|---|
| 0 | 130 | 40 |
| 0.1 | 128.59310831 | 42.67538466 |
| 0.4 | 125.29096506 | 48.3918314 |
| 0.8 | 116.74679409 | 61.0637458 |
| 1.2 | 105.89029352 | 74.64137076 |
| 1.6 | 93.43402157 | 87.80892587 |
| 2.0 | 80.39431734 | 99.05842237 |
| 2.4 | 67.81178199 | 107.14296048 |
| 2.8 | 56.47832272 | 111.43100139 |
| 3.2 | 46.81039171 | 111.97738337 |
| 3.6 | 38.88500361 | 109.34194238 |
| 4.0 | 32.55947271 | 104.31904479 |
| 4.4 | 27.59188441 | 97.71559988 |
| 4.8 | 23.72338715 | 90.22535562 |
| 5 | 22.12680352 | 86.32152735 |

all terms are considered, which is the most natural situation, the dynamics of Eq. (6.9) are given in Figure 6.3. From Figure 6.3, we have observed that uninfected people become infected when they came in contact with infected people. Similarly, the second numerical illustrative is solved by all three cases and dynamics is given in Figure 6.4, Figure 6.5, and Figure 6.6.

## 6.7 CONCLUSION

In this chapter, we have focused on the solving technique of the prey–predator model and its stability analysis. All the results are given in the form of a theorem with appropriate assumptions. By using Section 6.2.4, the closed-form solution Eq. (6.1) is obtained, which contains a periodic function and describes the cyclic behavior of both species (prey and predator). Stability analysis shows that this classical model behaves like an undamped oscillator with complex eigenvalues. The capability of the proposed technique is shown by solving the application and the example.

## REFERENCES

1. Letetia Addison, B. Bishwaroop, O. David, 2017. A financial prey–predator model with infection in predator, *Journal of Advances in Mathematics and Computer Science*, 25(6), 1–16.
2. W. Jinrong, F. Michel, 2020. Dynamics of a discrete nonlinear prey–predator model, *International Journal of Bifurcation and Chaos*, 30(4), 2050055.
3. Raw, S.N., Mishra, P., Tiwari, B. 2020. Mathematical study about a predator–prey model with anti-predator behavior, *International Journal of Applied and Computational Mathematics*, 6, 68.
4. Pandit P., Singh P. 2020. Fully fuzzy semi-linear dynamical system solved by fuzzy Laplace transform under modified Hukuhara derivative. In Ed. Das K., Bansal J., Deep K., Nagar A., Pathipooranam P., Naidu R. *(Soft Computing for Problem Solving) Advances in Intelligent Systems and Computing*, Springer, Singapore. Vol 1048.
5. Vijya Laxmi G.M, 2020. Effect of herd behaviour prey–predator model with competition in predator, *Materials Today Proceedings*, 33(7), 3197–3200, ISSN 2214-7853, https://doi.org/10.1016/j.matpr.2020.04.166.
6. A. Ahmed, N.K. Raid, 2018. A study of a diseased prey–predator model with refuge in prey and harvesting from predator, *Journal of Applied Mathematics*, 2, 1–17.
7. Pandit Purnima, Singh Payal, 2014. Prey–predator model and fuzzy initial condition, *International Journal of Engineering and Innovative Technology (IJEIT)*, 3(12), 65–68.
8. J. Gani, Randall J. Swift, 2013. Prey–predator model with infected prey and predators, *Discrete and Continuous Dynamical Systems*, 33(11), 5059–5066.
9. J. J. Buckley, L.J. Jowers, 2006. *Simulating Continuous Fuzzy Systems*, Springer, Berlin Heidelberg.

# 7 Incremental Harmonic Balance Method for Multi-Degree-of-Freedom System with Time-Delays

*R.K. Mitra*

## CONTENTS



## 7.1 INTRODUCTION

The complex behavior of nonlinear systems, such as periodic, aperiodic, and chaotic solutions, is generally dealt with by active vibration control, based on time-delayed feedback control. The delay in the feedback path is expected for several reasons or is sometimes introduced intentionally to obtain the desired performance of the nonlinear dynamical system. Generally, the presence of a delay, either natural or intentional, results in the more complex behavior of the dynamic system. Many researchers (Plaut and Hsieh, 1987a, b; Tsuda et al., 1992; Ji and Hansen, 2006) have investigated the complex dynamics and instability of the controlled systems caused by unavoidable delays in controllers and actuators.

From the literature, it is observed that vibration control of a certain class of nonlinear system was mostly investigated by applying perturbation techniques, numerical integration (NI) techniques, or a straightforward harmonic balance (HB) analysis. The major drawback of the methods was that they were not able to trace all possible solutions, both stable and unstable, or all possible bifurcations. Furthermore, the time integration technique for highly stiff equations is expensive and inefficient for parametric studies. In contrast, the harmonic balance method is only grossly approximate with lesser harmonics and needs reformulation each time the number of harmonics is changed. The present study is motivated by the need for a better semi-analytical prediction of complex periodic (ultra-subharmonic) nonlinear systems *via* the incremental harmonic balance method (IHB), as previous theoretical analysis focused on weakly nonlinear regimes and were mostly suitable for systems with small amplitudes of oscillation.

Recently, several works employing the IHB method have been reported. Bhattiprolu et al. (2016) studied the periodic response of beams on nonlinear and viscoelastic unilateral foundations. Nonlinear analysis of a parametrically excited beam with intermediate support has been studied by Zhou et al. (2016). Shen et al. (2016) studied the fractional order nonlinear oscillator, while Yuanping and Siyu (2016) investigated periodic responses and bifurcation behavior of a suspension system. The nonlinear vibrations of the bilinear hysteretic oscillator were investigated by Huai et al. (2017), and Liu et al. (2018) studied quasi-periodic aeroelastic responses of an airfoil with the external store. Wang et al. (2019) investigated applications of the IHB method combined with equivalent piecewise linearization on vibrations of nonlinear stiffness systems. Ranjbarzadeh and Kakavand (2019) examined nonlinear vibration of a two degree-of-freedom system with an asymmetric piecewise-linear compression spring. Nonreciprocal transmission of nonlinear elastic wave metamaterials was investigated by Wei et al. (2020), while Wang et al. (2020) examined periodic forced oscillation of a dielectric elastomer balloon, employing the IHB method.

Therefore, for dynamics and vibration control analysis of a system having strong nonlinearity with the feedback control law (introduced in the feedback path), a theoretical basis of the frequency domain-based computer method is necessary for the study of resonances in a systematic way that is both robust and better for convergence of the solution. To address these issues, an incremental harmonic balance, along with the continuation technique (IHBC), a systematic computer method, is applied for time-delayed feedback control of nonlinear systems under harmonic excitation.

In the present study, the incremental harmonic balance method and continuation technique (IHBC) have been developed to study the fundamental (period-1) and higher order harmonic frequency response curves for the damped Duffing oscillator with time-delayed displacement and velocity feedback. In the present study, the stability of uncontrolled responses is obtained by Floquet's theory, using Hsu's scheme, whereas the stability of controlled responses is obtained using the semi-discretization method for delay differential equations.

## 7.2 FORMULATION OF IHB METHOD FOR DELAY DIFFERENTIAL EQUATIONS

Oscillations of nonlinear dynamical systems, with $N$ degrees of freedom under time-delayed state (displacement and velocity) feedback and subjected to periodic excitation, can be described by a system of nonlinear delay differential equations (DDEs).

$$x_j'' + \varphi_j(x_1', \ldots, x_N', \; x_1, \ldots, x_1, \; x_{d1}', \ldots, x_{dN}', \; x_{d1}, \ldots, x_{dN}, \lambda, \tau) = F_j(h\omega\tau), \tag{7.1}$$

In this vector equation, $\phi_j$ denotes nonlinear functions of dependent variables $x_j(\tau)$. $x_j$ is the unknown response of the nonlinear system as a function of time $\tau$ (in seconds). An over dash denotes a derivative with respect to $\tau$. $\omega$ is the excitation frequency in rad/s and $d$ is the time delay in seconds. $F_j(h\omega\tau)$ is the harmonic excitation function, $h$ is an integer to consider subharmonic and superharmonic responses, and $h \in \{1,2,\ldots\}$. $x_{dj}(\tau) = x_j(\tau - d)$ and $x_{dj}'(\tau) = x_j'(\tau - d)$ are the control signals due to time-delayed displacement and velocity feedback, respectively. Here, the responses $x_j(\tau)$, $x_j'(\tau)$ and $x_j''(\tau)$ are time-periodic, i.e.,

$$x_j(\tau) = x_j(\tau + 2h\pi), \;\; x_j'(\tau) = x_j'(\tau + 2h\pi) \text{ and } x_j''(\tau) = x_j''(\tau + 2h\pi). \; \left(j = 1 \text{ to } N\right) \tag{7.2}$$

Introducing dimensionless time $t = \omega\tau$ as an independent variable, the dynamical equation, amenable to the IHB form, expressed as Eq. (7.1), may be written as:

$$\begin{aligned} &\omega^2 \ddot{x}_j + \phi_j(\dot{x}_1, \ldots, \dot{x}_N, \; x_1, \ldots, x_1, \; \dot{x}_{d1}, \ldots, \dot{x}_{dN}, \\ &x_{d1}, \ldots, x_{dN}, \omega, \lambda, t) = F_j(ht), \; \left(j = 1 \text{ to } N\right) \end{aligned} \tag{7.3}$$

Here, an over dot denotes a derivative with respect to $t$. In vector Eq. (7.3), $\varphi_j$ denotes nonlinear functions of dependent variables $x_j$ and corresponding first derivatives $\dot{x}_j = \frac{1}{\omega}\frac{dx_j}{d\tau}$, second derivatives $\ddot{x}_j = \frac{1}{\omega}\frac{d^2x_j}{d\tau^2}$, the exciting frequency $\omega$, the time delay parameter $d$ , and the variable parameter $\lambda$. $F_j(ht)$ is the harmonic excitation function, and $x_{dj}(t) = x_j(t - \omega d)$ and $\dot{x}_{dj}(t) = \dot{x}_j(t - \omega d)$ are the control signals due to time-delayed displacement and velocity feedback, respectively.

The steady-state solution of Eq. (7.3) is periodic and assumed to be in the form of a Fourier series expressed as:

$$x_j(t) = a_{j0} + \sum_{i=1}^{n} \left[ a_{ji} \cos\left(it\right) + b_{ji} \sin\left(it\right) \right], \tag{7.4}$$

where $a_{j0}$, $a_{ji}$ and $b_{ji}$ ($i = 1$ to $n$) are the Fourier coefficients for steady-state response $x_j(t)$.

Eq. (7.4) can be written in the matrix form as

$$\{x\} = [Y]\{A\}, \tag{7.5}$$

where $[Y]$ is a row matrix of dimension $1 \times (2n+1)$ and $\{A\}$ is a column matrix of dimension $(2n+1) \times 1$ and expressed as:

$$[Y]=\begin{bmatrix}1 & \cos t & \cos 2t & \dots & \cos nt & \sin t & \sin 2t & \dots & \sin nt\end{bmatrix}, \quad (7.6)$$

and

$$\{A\}=\begin{bmatrix}a_{j0} & a_{j1} & a_{j2} & \dots & a_{jn} & b_{j1} & b_{j2} & \dots & b_{jn}\end{bmatrix}^{\mathrm{T}}, \quad (7.7)$$

where 'T' denotes the transpose of a matrix.

The time-delayed displacement and velocity are expressed as:

$$x_{dj}=x_j(t-\omega d) \text{ and } \dot{x}_{dj}=\dot{x}_j(t-\omega d). \quad (7.8)$$

In the form of a Fourier series, the delayed terms are expressed as:

$$x_{dj}=a_{j0}+\sum_{i=1}^{n}\left[a_{ji}\cos\{i(t-\omega d)\}+b_{ji}\sin\{i(t-\omega d)\}\right]=[Y_d]\{A_j\}, \quad (7.9)$$

$$\dot{x}_{dj}=\sum_{i=1}^{n}\left[-(i)a_{ji}\sin\{i(t-\omega d)\}+(i)b_{ji}\cos\{i(t-\omega d)\}\right]=[\dot{Y}_d]\{A_j\}, \quad (7.10)$$

where $[Y_d]=[1 \quad \cos(t-\omega d) \quad \cos 2(t-\omega d) \quad \dots \quad \cos n(t-\omega d)$

$$\sin(t-\omega d) \quad \sin 2(t-\omega d) \quad \dots \quad \sin n(t-\omega d)\big], \quad (7.11)$$

and $[\dot{Y}_d]=[0 \quad -\sin(t-\omega d) \quad -2\sin 2(t-\omega d) \quad \dots \quad -n\sin n(t-\omega d)$

$$\cos(t-\omega d) \quad 2\cos 2(t-\omega d) \quad \dots \quad n\cos n(t-\omega d)\big]. \quad (7.12)$$

The first step of the IHB method is a Newton-Raphson iterative procedure. To obtain a periodic solution of Eq. (7.3), one needs to guess a solution at the beginning of the procedure, which may be taken as the solution of the corresponding linear system. Assuming that some initial solution or initial guess determining the initial state is known, one tries to obtain a neighboring solution by adding the corresponding increments (symbolized by $\Delta$) to them as follows:

$$x_j \to x_j+\Delta x_j, \quad \omega \to \omega+\Delta\omega \text{ and } \lambda \to \lambda+\Delta\lambda, \quad (7.13)$$

$$x_{dj} \to x_{dj}+\Delta x_{dj}, \text{ or } x_j\left(t-\omega d\right) \to x_j\left(t-\omega d\right)+\Delta x_j\left(t-\omega d\right). \quad (7.14)$$

It is worth noting that the first terms on the right-hand side of Eqs. (7.13) and (7.14) are known. For higher-order derivatives of $x_j$ and $x_{dj}$, we can write

$$\dot{x}_j \to \dot{x}_j+\Delta\dot{x}_j, \ \ddot{x}_j \to \ddot{x}_j+\Delta\ddot{x}_j, \quad (7.15)$$

$$\dot{x}_{dj} \to \dot{x}_{dj} + \Delta\dot{x}_{dj}, \text{ or } \dot{x}_j(t-\omega d) \to \dot{x}_j(t-\omega d) + \Delta\dot{x}_j(t-\omega d). \tag{7.16}$$

Substituting these terms in Eq.(7.2), we obtain

$$(\omega+\Delta\omega)^2(\ddot{x}_j+\Delta\ddot{x}_j)+\varphi_j(\dot{x}_j+\Delta\dot{x}_j, x_j+\Delta x_j, \dot{x}_{dj}+\Delta\dot{x}_{dj}, x_{dj}+\Delta x_{dj}),$$

$$\omega+\Delta\omega,\ \lambda+\Delta\lambda,\ t) = F_j(ht). \tag{7.17}$$

where $j=1$ to $N$. The increment of $x_j$ is expressed by

$$\Delta x_j = \Delta a_{j0} + \sum_{i=1}^{n}\left[\Delta a_{ji}\cos(it)+\Delta b_{ji}\sin(it)\right] = [Y]\{\Delta A_j\}, \tag{7.18}$$

$$\text{where } \{\Delta A_j\} = \begin{bmatrix}\Delta a_{j0} & \Delta a_{j1} & \Delta a_{j2} & \dots & \Delta a_{jn} & \Delta b_{j1} & \Delta b_{j2} & \dots & \Delta b_{jn}\end{bmatrix}^{\mathrm{T}}. \tag{7.19}$$

Similarly delayed increments are

$$\Delta x_{dj} = [Y_d]\{\Delta A_j\} \text{ and } \Delta\dot{x}_{dj} = [\dot{Y}_d]\{\Delta A_j\}. \tag{7.20}$$

Expanding Eq. (7.8) by Taylor's series about the initial state and ignoring all higher-order terms of small increments, the linearized incremental equations are obtained as:

$$\begin{aligned}&\omega^2\Delta\ddot{x}_j + \sum_{k=1}^{N}\left(\frac{\partial\phi_j}{\partial\dot{x}_k}\right)\Delta\dot{x}_j + \sum_{k=1}^{N}\left(\frac{\partial\phi_j}{\partial x_k}\right)\Delta x_j + \sum_{k=1}^{N}\left(\frac{\partial\phi_j}{\partial\dot{x}_{dk}}\right)\Delta\dot{x}_{dj} + \sum_{k=1}^{N}\left(\frac{\partial\phi_j}{\partial x_{dk}}\right)\Delta x_{dj}\\ &+\left(\frac{\partial\phi_j}{\partial\omega}+2\omega\ddot{x}_j\right)\Delta\omega + \left(\frac{\partial\phi_j}{\partial\lambda}\right)\Delta\lambda = F_j(ht)-\omega^2\ddot{x}_j-\phi_j\end{aligned} \tag{7.21}$$

The initial values and increments are collected in vector forms

$$\{\mathbf{X}\} = \begin{bmatrix}x_1 & x_2 & \dots & x_N\end{bmatrix}^T,\quad \{\Delta\mathbf{X}\} = \begin{bmatrix}\Delta x_1 & \Delta x_2 & \dots & \Delta x_N\end{bmatrix}^T, \tag{7.22}$$

$$\{\mathbf{X}_d\} = \begin{bmatrix}x_{d1} & x_{d2} & \dots & x_{dN}\end{bmatrix}^T,\quad \{\dot{\mathbf{X}}_d\} = \begin{bmatrix}\dot{x}_{d1} & \dot{x}_{d2} & \dots & \dot{x}_{dN}\end{bmatrix}^T, \tag{7.23}$$

and

$$\{\mathbf{F}\} = \begin{bmatrix}F_1 & F_2 & \dots & F_N\end{bmatrix}^T. \tag{7.24a}$$

Let us define matrices [**C**], [**K**], [**D**] and $\left[\bar{D}\right]$ with the corresponding elements

$$C_{jk}=\left(\frac{\partial\phi_j}{\partial\dot{x}_k}\right),\ K_{jk}=\left(\frac{\partial\phi_j}{\partial x_k}\right),\ D_{jk}=\left(\frac{\partial\phi_j}{\partial x_{dk}}\right)\text{ and }\bar{D}_{jk}=\left(\frac{\partial\phi_j}{\partial\dot{x}_{dk}}\right). \tag{7.24b}$$

and vectors $\{\mathbf{\Omega}\}$, $\{\hat{\mathbf{\Omega}}\}$, and $\{\mathbf{R}\}$ with the corresponding elements

$$\Omega_j=\left(\frac{\partial\phi_j}{\partial\omega}\right),\ \hat{\Omega}_j=\left(\frac{\partial\phi_j}{\partial\lambda}\right)\text{ and }R_j=F_j(ht)-\omega^2\ddot{x}_j-\phi_j. \tag{7.25}$$

The linearized incremental equations can be written in the convenient matrix form as:

$$\begin{aligned}&\omega^2\{\Delta\ddot{\mathbf{X}}\}+[\mathbf{C}]\{\Delta\dot{\mathbf{X}}\}+[\mathbf{K}]\{\Delta\mathbf{X}\}+[\mathbf{D}]\{\Delta\mathbf{X_d}\}+[\bar{\mathbf{D}}]\{\Delta\dot{\mathbf{X}}_\mathbf{d}\}\\&+\left(\{\mathbf{\Omega}\}+2\omega\{\ddot{\mathbf{X}}\}\right)\Delta\omega+\{\hat{\mathbf{\Omega}}\}\Delta\lambda=\{\mathbf{R}\}.\end{aligned} \tag{7.26}$$

Additionally, the matrices $[\mathbf{Y}]$, $[\mathbf{Y_d}]$, $[\dot{\mathbf{Y}}_\mathbf{d}]$ and vectors $\{\mathbf{A}\}$ and $\{\Delta\mathbf{A}\}$ are introduced in the following forms:

$$[\mathbf{Y}]=\begin{bmatrix}[Y]&0&\cdots&0\\0&[Y]&\cdots&0\\\vdots&\vdots&\vdots&\vdots\\0&0&\cdots&[Y]\end{bmatrix},\ [\mathbf{Y_d}]=\begin{bmatrix}[Y_d]&0&\cdots&0\\0&[Y_d]&\cdots&0\\\vdots&\vdots&\vdots&\vdots\\0&0&\cdots&[Y_d]\end{bmatrix}, \tag{7.27}$$

$$[\dot{\mathbf{Y}}_\mathbf{d}]=\begin{bmatrix}[\dot{Y}_d]&0&\cdots&0\\0&[\dot{Y}_d]&\cdots&0\\\vdots&\vdots&\vdots&\vdots\\0&0&\cdots&[\dot{Y}_d]\end{bmatrix},\ \{\mathbf{A}\}=\begin{bmatrix}\{A_1\}\\\{A_2\}\\\vdots\\\{A_N\}\end{bmatrix},\ \{\Delta\mathbf{A}\}=\begin{bmatrix}\{\Delta A_1\}\\\{\Delta A_2\}\\\vdots\\\{\Delta A_N\}\end{bmatrix}. \tag{7.28}$$

Eqs. (7.27) and (7.28) are used in the interpretation of the following vectors:

$$\{\mathbf{X}\}=[\mathbf{Y}]\{\mathbf{A}\},\ \{\mathbf{X_d}\}=[\mathbf{Y_d}]\{\mathbf{A}\}\text{ and }\{\dot{\mathbf{X}}_\mathbf{d}\}=[\dot{\mathbf{Y}}_\mathbf{d}]\{\mathbf{A}\}, \tag{7.29}$$

and the corresponding increments

$$\{\Delta\mathbf{X}\}=[\mathbf{Y}]\{\Delta\mathbf{A}\},\ \{\Delta\mathbf{X_d}\}=[\mathbf{Y_d}]\{\Delta\mathbf{A}\}\text{ and }\{\Delta\dot{\mathbf{X}}_\mathbf{d}\}=[\dot{\mathbf{Y}}_\mathbf{d}]\{\Delta\mathbf{A}\}. \tag{7.30}$$

Replacing the matrices of Eqs. (7.29) and (7.30) in Eq. (7.26), we obtain

$$\begin{aligned}&\left(\omega^2[\ddot{\mathbf{Y}}]+[\mathbf{C}][\dot{\mathbf{Y}}]+[\mathbf{K}][\mathbf{Y}]+[\mathbf{D}][\mathbf{Y_d}]+[\bar{\mathbf{D}}][\dot{\mathbf{Y}}_\mathbf{d}]\right)\{\Delta\mathbf{A}\}\\&+\left(\{\mathbf{\Omega}\}+2\omega[\ddot{\mathbf{Y}}]\{\mathbf{A}\}\right)\Delta\omega+\{\hat{\mathbf{\Omega}}\}\Delta\lambda=\{\mathbf{R}\}\end{aligned} \tag{7.31}$$

Matrix Eq. (7.31) is linear with variable coefficients and can be solved by Galerkin's procedure, which is the second step in the IHB method.

Applying Galerkin's procedure for one time-period of response, we obtain

$$\begin{aligned}&\left[\int_0^{2h\pi}\left[\mathbf{Y}\right]^{\mathrm{T}}\left(\omega^2\left[\ddot{\mathbf{Y}}\right]+\left[\mathbf{C}\right]\left[\dot{\mathbf{Y}}\right]+\left[\mathbf{K}\right]\left[\mathbf{Y}\right]+\left[\mathbf{D}\right]\left[\mathbf{Y_d}\right]+\left[\bar{\mathbf{D}}\right]\left[\dot{\mathbf{Y}}_\mathbf{d}\right]\right)dt\right]\{\Delta\mathbf{A}\}\\&+\left[\int_0^{2h\pi}\left[\mathbf{Y}\right]^{\mathrm{T}}\left(\{\Omega\}+2\omega\left[\ddot{\mathbf{Y}}\right]\{\mathbf{A}\}\right)dt\right]\Delta\omega+\left[\int_0^{2h\pi}\left(\{\hat{\Omega}\}\right)dt\right]\Delta\lambda=\left[\int_0^{2h\pi}\{\mathbf{R}\}\,dt\right].\end{aligned}\tag{7.32}$$

Replacing

$$\left[\Pi\right]=\int_0^{2h\pi}\left[\mathbf{Y}\right]^{\mathrm{T}}\left(\omega^2\left[\ddot{\mathbf{Y}}\right]+\left[\mathbf{C}\right]\left[\dot{\mathbf{Y}}\right]+\left[\mathbf{K}\right]\left[\mathbf{Y}\right]+\left[\mathbf{D}\right]\left[\mathbf{Y_d}\right]+\left[\bar{\mathbf{D}}\right]\left[\dot{\mathbf{Y}}_\mathbf{d}\right]\right)dt,\tag{7.33}$$

$$\{\Theta\}=\int_0^{2h\pi}\left[\mathbf{Y}\right]^{T}\left(\{\Omega\}+2\omega\left[\ddot{\mathbf{Y}}\right]\{\mathbf{A}\}\right)dt,\tag{7.34}$$

$$\{\hat{\Theta}\}=\int_0^{2h\pi}\left[\mathbf{Y}\right]^{\mathrm{T}}\{\hat{\Omega}\}\,dt\tag{7.35}$$

and

$$\{\Phi\}=\int_0^{2h\pi}\left[\mathbf{Y}\right]^{\mathrm{T}}\{\mathbf{R}\}\,dt,\tag{7.36}$$

Eq. (7.31) is expressed as:

$$\left[\Pi\right]\{\Delta\mathbf{A}\}+\{\Theta\}\Delta\omega+\{\hat{\Theta}\}\Delta\lambda=\{\Phi\}.\tag{7.37}$$

where the matrix $[\boldsymbol{\Pi}]$ has a dimension $(2n+1)N\times(2n+1)N$ and vectors $\{\boldsymbol{\Theta}\}$, $\{\hat{\Theta}\}$, and $\{\boldsymbol{\Phi}\}$ have the dimension $(2n+1)N\times 1$.

Eq. (7.37) is a linear matrix equation for the unknown vector of harmonic increments $\{\Delta\mathbf{A}\}$. $[\boldsymbol{\Pi}]$ is known as the Jacobian (gradient or tangential) matrix with respect to $\{\Delta\mathbf{A}\}$.

Finally, the matrix Eq. (7.37) is solved at each time-step of the Newton-Raphson iterative procedure with updated $[\boldsymbol{\Pi}]$ and $\{\boldsymbol{\Phi}\}$ until a convergent solution is obtained. With the assumption that $\Delta\omega$ vanishes and $\{\Delta\mathbf{A}\}$ represents an exact solution, the residual vector $\{\boldsymbol{\Phi}\}$ disappears and Eq. (7.37) guarantees that $\{\Delta\mathbf{A}\}$ vanishes. $\{\mathbf{A}\}$

is an approximation, so the better solution, $\{\mathbf{A}\}+\{\Delta\mathbf{A}\}$, is computed by solving Eq. (7.37). This procedure is repeated until $\{\mathbf{\Phi}\}$ becomes sufficiently small.

After obtaining a solution for a particular control parameter, $\lambda$, it is then augmented by an increment $\Delta\lambda$ (known as the control increment). This is known as the augmentation process. The iteration process is again repeated using the previous solution as an approximation until the solution for the neighboring $\lambda$ is obtained. By successive use of augmentation and the iteration process, a complete solution diagram may readily be traced. The choice of control increment depends on the rate of change of the variation of the increments; usually, the one with the greatest increment (absolute values) will be chosen as the control increment. This is extremely useful when a solution diagram is very sensitive to the variation in the parameter $\omega$ or similar, and the number of steps to trace the solution diagram has to be reduced.

To be more efficient in tracing the solution diagram, an arc-length continuation technique can be used to predict the next solution from some known states. With such a prediction, the number of iterations required for a convergent solution will be greatly reduced. Furthermore, it is also useful for solving the convergence problem at peaks of the solution diagram. Along an equilibrium path for a varying parameter, it is possible that the solution curve may form a loop (when the curve cuts itself again) and a knot (when the curve does not cut into itself again). In order to eliminate the singular behavior of $\lambda$ at folds, one may use either the selective coefficient method or the arc-length method. The latter method is used in the present study and discussed in Section 7.3.

## 7.3 PATH-FOLLOWING AND PARAMETRIC CONTINUATION

The IHB method with a variable parameter is ideally suited to parametric continuation to obtain the response diagrams of nonlinear systems with time-delayed feedback. The solution diagram for the parametric study can be obtained with the arc-length method, and the IHB technique can be comprehensively infused in the same program. The resulting procedure for obtaining a solution diagram is known as IHBC. After obtaining the solution for the particular value of a parameter, the solution for another parameter perturbed from the original one can be obtained by iterations, using the previous solution as an approximation. Both stable and unstable solutions will mix together. The stability of steady-state responses is obtained by Floquet's theory, which will be discussed in Section 7.4. As stable and unstable branches of the solution are traced by using a path-following technique with varying frequency parameter (as here), the eigenvalues of the monodromy matrix start moving and cross the unit circle, achieving different bifurcations in the system (Figure 7.1). When an eigenvalue crosses the (+1, 0) boundary of the unit circle, a fold bifurcation occurs when a stable and an unstable saddle attractor coalesce. It is also known as the saddle-node bifurcation and is associated with a jump phenomenon. When an eigenvalue crosses the (–1, 0) boundary of the unit circle, a flip bifurcation occurs. This is related to the period-doubling bifurcation. When a pair of complex conjugate eigenvalues leave the unit circle (radially), a secondary Hopf or Neimark bifurcation occurs. A Hopf bifurcation introduces a new frequency, possibly not corresponding

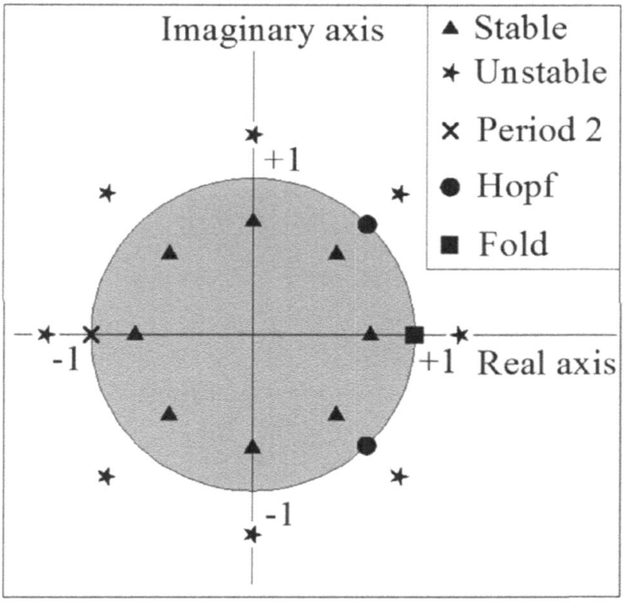


**FIGURE 7.1** Stability criteria depend on real and imaginary parts of eigenvalues of the transition matrix. If all the absolute values of eigenvalues remain within the unit circle (gray shading), the solution is stable, otherwise it is unstable. The manner in which the eigenvalues leave the unit circle produces different bifurcation behavior, as shown.

to the first one in the bifurcation solution. The bifurcation solution may be periodic or quasi-periodic, depending on the relationship between the newly introduced frequency and the frequency of the periodic solution that existed before the bifurcation. When an eigenvalue touches the (+1, 0) boundary of the unit circle from the inside and travels backward, that is, toward the origin, a symmetry-breaking bifurcation occurs, in which a periodic solution with only odd harmonic coefficients bifurcates to the periodic solution, with odd and even harmonic coefficients.

The main aim of the path-following and the parametric continuation is to effectively trace the bifurcation sequence as a parameter of the system is varied. In this study, an arc- length procedure (Leung and Chui 1995) is adopted for the parametric continuation and developed for the present study with time-delay terms. Introducing the path parameter, $\gamma$, the augmenting equation for a general system can be written as:

$$\psi(\chi)-\gamma=0, \tag{7.38}$$

where

$$\chi=\left[\{\mathbf{A}\},\lambda\right]^{\mathrm{T}}.$$

$$From\ Eq.\ (7.28), \{\mathbf{A}\} = \begin{bmatrix} \{A_1\} \\ \{A_2\} \\ \vdots \\ \{A_N\} \end{bmatrix}, \tag{7.39}$$

and

$$\{A_j\} = \begin{bmatrix} a_{j0} & a_{j1} & a_{j2} & \dots & a_{jn} & b_{j1} & b_{j2} & \dots & b_{jn} \end{bmatrix}^{\mathrm{T}}, \ (j = 1 \text{ to } N). \tag{7.40}$$

A good choice for the function $\psi(\chi)$ is $\psi = \chi^{\mathrm{T}}\chi$. Taking increments on $\{\mathbf{A}\}$, $\lambda$, and $\gamma$, the incremental equation is obtained as:

$$\sum_{i=1}^{(2n+1)N} \frac{\partial \psi}{\partial(\{A_i\})}\{\Delta A_i\} + \frac{\partial \psi}{\partial \lambda}\Delta\lambda - \Delta\gamma + \psi(\chi) - \gamma = 0, \tag{7.41}$$

where $\{A_i\}$ is the $i$th element of $\{\mathbf{A}\}$. Together with Eq. (7.19), the modified augmented equation is expressed by

$$\begin{bmatrix} [\Pi] & \{\widehat{\Theta}\} \\ \left\{\dfrac{\partial \psi}{\partial\{\mathbf{A}\}}\right\}^{\mathrm{T}} & \left\{\dfrac{\partial \psi}{\partial \lambda}\right\}^{\mathrm{T}} \end{bmatrix} \begin{Bmatrix} \{\Delta\mathbf{A}\} \\ \Delta\lambda \end{Bmatrix} = \begin{Bmatrix} \{\Phi\} \\ \gamma + \Delta\gamma - \psi \end{Bmatrix}, \tag{7.42}$$

where the modified Jacobian matrix, having a dimension $(2n+2)N \times (2n+2)N$, is equal to

$$[\Pi] = \begin{bmatrix} [\Pi] & \{\widehat{\Theta}\} \\ \left\{\dfrac{\partial \psi}{\partial\{\mathbf{A}\}}\right\}^{\mathrm{T}} & \left\{\dfrac{\partial \psi}{\partial \lambda}\right\}^{\mathrm{T}} \end{bmatrix}, \tag{7.43}$$

Considering the portion of the equilibrium path of the solution branches, as shown in Figure 7.2, the augmenting equation can be written as

$$\psi(\chi) - \gamma - \{\chi'\}^T\{\chi - \chi_C\} = 0. \tag{7.44}$$

The first prediction of the new point $\{\chi_{New}\}$ of the solution along the equilibrium path is given in terms of the two previous points, $\{\chi_C\}$ and $\{\chi_{CC}\}$, as follows:

$$\left.\begin{aligned} \{\chi_{New}\} &= \{\chi_C\} + \Delta\gamma\{\chi'\}, \quad \text{and} \\ \{\chi'\} &= \{\chi_C - \chi_{CC}\} / \| \chi_C - \chi_{CC} \| \end{aligned}\right\}, \tag{7.45}$$

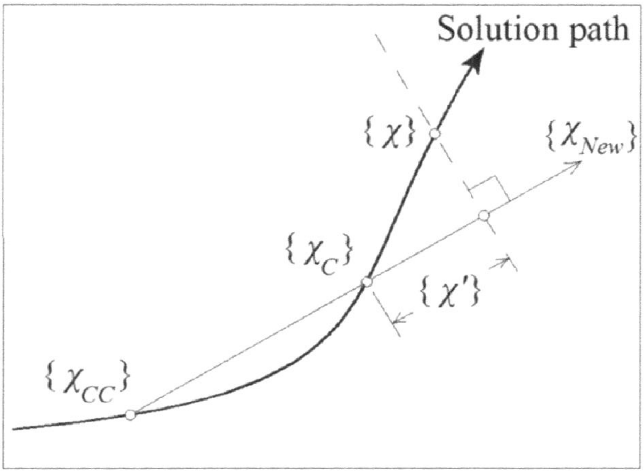


**FIGURE 7.2** A portion of the equilibrium path where the prediction of a new solution $\{\chi_{New}\}$ is obtained with the help of two previously known solutions $\{\chi_{CC}\}$ and $\{\chi_C\}$.

where $\|\chi_C - \chi_{CC}\|$ implies the norm of the matrix $(\chi_C - \chi_{CC})$, and $\Delta\gamma$ is an arbitrary step length taken in the computation from experience.

In certain cases, especially for a time-delayed feedback system with high gains and delays, the solution may approach limit points, resulting in failure of convergence. In such a case, scaling of parameters may aid in the convergence of solutions (Seydel, 1988). For the present study (Figure 7.3), the scaling factor can be calculated as:

$$\sigma = \left(\frac{1}{\omega_L}\right)\left(\sum_{i=1}^{N(2n+1)} \chi_{Li}^2\right)^{1/2}, \tag{7.46}$$

and the new step length, $\Delta\bar{\gamma}$, is calculated to be

$$\Delta\bar{\gamma}^2 = \left(\omega_L - \omega_C\right)^2 + \sum_{i=1}^{N(2n+1)} \left(\frac{\chi_{Li} - \chi_{Ci}}{\sigma}\right)^2, \tag{7.47}$$

where $\chi_{Ci}$ are the components of $\{\chi_C\}$, which refers to the previously converged solution and $\omega_C$ is the previously converged frequency. $\chi_{Li}$ and $\omega_L$ refer to the latest estimations of the solution.

## 7.4 STABILITY ANALYSES OF PERIODIC SOLUTIONS

In the IHB method, the periodic solution is assumed to be a Fourier series, the coefficients of which are independent of time. Therefore, it can provide both stable and unstable solutions. For static problems, changes in the stability of a solution occur at singular points characterized by folds and pitchforks, which can be detected by the

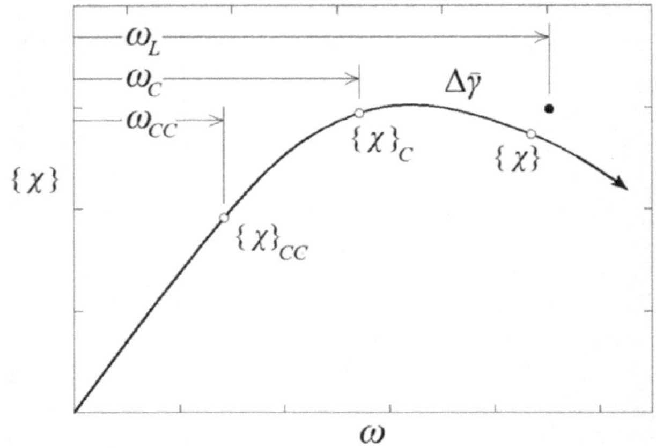


**FIGURE 7.3** A solution trajectory to estimate modified step length ($\Delta\bar{\gamma}$) based on scaling factor ($\sigma$).

disappearance of det [Π]. The methods cannot be applied to dynamic problems in which change instability can also occur at singular points characterized by period-doubling and Hopf bifurcations. For period-doubling, the number of terms in the Fourier expansion equation, Eq. (7.4), must be doubled by increasing $n$ to $2n$, with the determinant of the resulting Jacobian matrix of a double order becoming zero. If the original subharmonic of order is considered, the Floquet method is capable of detecting the period-doubling points. The periodic (or almost periodic) solution is checked for stability by means of the Floquet theory for two important reasons. Firstly, stable branches can be distinguished from unstable ones, and, secondly, bifurcation points can be located by monitoring eigenvalues of the monodromy matrix.

In the present study, the stability of uncontrolled responses is obtained by Floquet's theory, using Hsu's scheme (Hsu, 1972; 1974; Hsu and Cheng, 1973), whereas the stability of controlled responses is obtained using the semi-discretization method for delay differential equations.

### 7.4.1 Floquet's Theory for an Uncontrolled System, Using Hsu's Scheme

In the context of the IHBC, the local stability analysis of the steady-state solution can be easily performed by perturbing the state variables in the neighborhood of the computed or given solution. This results in a system of linearized equations with periodically varying coefficients. The perturbed equation of motion is always autonomous. If the forcing frequency, as well as the amplitude of the forcing function, is not changed in Eq. (7.26), a small change in the solution, from $\{\mathbf{X}\}$ to $\{\mathbf{X}\}+\{\Delta\mathbf{X}\}$ will satisfy the following perturbed equation for the uncontrolled system:

$$\omega^2\{\Delta\ddot{\mathbf{X}}\}+[\mathbf{C}]\{\Delta\dot{\mathbf{X}}\}+[\mathbf{K}]\{\Delta\mathbf{X}\}=\{\mathbf{0}\}, \tag{7.48}$$

which is a matrix ordinary differential equation with periodic coefficients in matrices [**C**] and [**K**]. The corresponding elements of the aforesaid matrices are written again to maintain clarity.

$$C_{jk} = \left( \frac{\partial \phi_j}{\partial \dot{x}_k} \right) \text{ and } K_{jk} = \left( \frac{\partial \phi_j}{\partial x_k} \right). \tag{7.49}$$

It is worth mentioning that all the elements need to be evaluated at the known solution {**X**}. The state-space form of Eq. (7.48) is:

$$\frac{d}{dt}\begin{bmatrix} \{\Delta \mathbf{X}\} \\ \{\Delta \dot{\mathbf{X}}\} \end{bmatrix} = -\frac{1}{\omega^2}\begin{bmatrix} [\mathbf{0}] & [\mathbf{I}] \\ [\mathbf{K}] & [\mathbf{C}] \end{bmatrix}\begin{bmatrix} \{\Delta \mathbf{X}\} \\ \{\Delta \dot{\mathbf{X}}\} \end{bmatrix}. \tag{7.50}$$

$$\text{Alternatively, } \{\dot{\mathbf{Z}}\} = [\Gamma]\{\mathbf{Z}\}, \tag{7.51}$$

where the transition matrix [**Γ**] is periodic with the time period $T$, i.e., [**Γ**($t$)] = [**Γ**($t+T$)].

Based upon the Floquet-Lyapunov theorem, the transition matrix for a periodic system [**Γ**($t$)] = [**Γ**($t+T$)] can be expressed by:

$$[\kappa(t, t_0)] = [\alpha(t)]^{-1}\{\exp[\beta(t - t_0)]\}[\alpha(t_0)], \tag{7.52}$$

where $[\boldsymbol{\alpha}(t+T)] = [\boldsymbol{\alpha}(t)]$ and $[\boldsymbol{\beta}]$ is a constant matrix. Clearly, the matrix determines the stability of the system $[\boldsymbol{\beta}]$, where $[\boldsymbol{\beta}]$ is given by the following relationship:

$$[\kappa(T, 0)] = \exp\left([\beta]T\right). \tag{7.53}$$

It is important to note that the Floquet-Lyapunov theorem does not yield the solution to the problem. However, it provides extremely valuable information about the form and properties of the solution. As a direct result of the Floquet-Lyapunov theorem, the knowledge of the transition matrix over a period determines the solution to the homogeneous system everywhere through the relationship

$$[\kappa(t + nT, 0)] = [\kappa(t, 0)]\{\exp([\beta]T)\}^n, \tag{7.54}$$

where $0 \le t \le T$ and $n$ is an integer.

It $[\boldsymbol{\beta}]$ has $n$ distinct eigenvalues, and a similarity transformation can be found such that

$$[\mu]^{-1}[\beta][\mu] = [\eta], \tag{7.55}$$

where the columns of $[\boldsymbol{\mu}]$ are the $n$-linearly independent eigenvectors of $[\boldsymbol{\beta}]$ and $[\boldsymbol{\eta}]$ is a diagonal matrix containing the eigenvalues of $[\boldsymbol{\beta}]$. Combining Eqs. (7.54) and (7.55) and using the definition of the matrix exponential, one obtains

$$\exp([\eta]T) = [\Lambda] = [\mu]^{-1}[\kappa(T,0)][\mu], \tag{7.56}$$

where $[\Lambda]$ is a diagonal matrix, containing the eigenvalues of the transition matrix at the end of one period, $T$. The eigenvalues of $[\kappa(T, 0)]$ (called characteristic multipliers) are related to the eigenvalues of $[\beta]$ (called characteristic exponents) through the relationship

$$\exp\left([\eta_k T]\right) = \left[\Lambda_k\right], \quad k = 1,\ 2,\ \ldots\ n. \tag{7.57}$$

Clearly, $\eta_k$ and $\Lambda_k$ are both complex quantities in general, and thus

$$\varsigma_k = \frac{1}{2T}\left\{\ln\left[\left(\Lambda_k\big|_{\text{Real}}\right)^2 + \left(\Lambda_k\big|_{\text{Imag}}\right)^2\right]\right\} \tag{7.58}$$

$$\varsigma_k = \frac{1}{2T}\left\{\tan^{-1}\left[\left(\Lambda_k\big|_{\text{Real}}\right)\Big/\left(\Lambda_k\big|_{\text{Imag}}\right)\right]\right\} \tag{7.59}$$

The quantity $\zeta_k$ can be determined according to the Floquet-Lyapunov theory only within an integer multiple of the non-dimensional time period $T$.

The stability criteria for the system are related to the eigenvalues of $[\beta]$ or the real part of the characteristic exponents $\zeta_k$. The solution of Eq. (7.48) approaches zero as $t \to \infty$ if

$$\left[\left\{\text{Real}\left(\Lambda_k\right)\right\}^2 + \left\{\text{Imag}\left(\Lambda_k\right)\right\}^2\right] < 1 \quad \text{or} \quad \varsigma_k < 0, \quad k = 1,2,\ldots n. \tag{7.60a}$$

For the general case, when the eigenvalues $[\beta]$ are not distinct, a similarity transformation can be used, which transforms $[\beta]$ into the Jordan canonical form. The efficient numerical method for evaluating the transition matrix at the end of one period $[\kappa(T, 0)]$ is presented in Section 7.4.2.

In a series of papers, Hsu et al. (Hsu, 1972; 1974; Hsu and Cheng, 1973) developed various methods for approximating the transition matrix during one period, of which the most efficient one consists of approximating the periodic matrix $[\Gamma(t)]$ by a series of step functions based on a generalization of the well-known rectangular ripple method applied to multi-dimensional systems.

### 7.4.2 Floquet's Theory for a Time-Delay System by the Semi-Discretization Method

Semi-discretization is a robust and powerful tool by which to investigate the stability behavior of delay differential equations (DDEs). If the forcing frequency, as well as the amplitude of the forcing function, is not changed in Eq. (7.1), a small change of solution from $\{\mathbf{X}\}$ to $\{\mathbf{X}\} + \{\Delta\mathbf{X}\}$ and $\{\mathbf{X_d}\}$ to $\{\mathbf{X_d}\} + \{\Delta\mathbf{X_d}\}$ will satisfy the following perturbed equation for the system with a time delay.

Here, $\{\mathbf{X}\} = [x_1\ x_2\ \ldots\ x_N]^T$ represents a column matrix of responses of an $N$ degree-of-freedom system and its increment is expressed by $\{\Delta\mathbf{X}\}$. $\{\mathbf{X_d}\}$ and $\{\dot{\mathbf{X}}_\mathbf{d}\}$ denote column matrices of time-delayed displacements and velocities, respectively. These two terms may be expressed as:

$$\{\mathbf{X_d}\} = \left[x_{d1}\ \ x_{d2} \ldots x_{dN}\right]^T, \tag{7.60b}$$

$$\{\dot{\mathbf{X}}_\mathbf{d}\} = \left[\dot{x}_{d1}\ \ \dot{x}_{d2} \ldots \dot{x}_{dN}\right]^T, \tag{7.61}$$

where $x_{dj} = x_j(t - \omega d)$ and $\dot{x}_{dj} = \dot{x}_j(t - \omega d)$, ($j = 1$ to $N$).

$$\omega^2\{\Delta\ddot{\mathbf{X}}\} + [\mathbf{C}]\{\Delta\dot{\mathbf{X}}\} + [\mathbf{K}]\{\Delta\mathbf{X}\} + [\mathbf{D}]\{\Delta\mathbf{X}_\mathrm{d}\} + [\bar{\mathbf{D}}]\{\Delta\dot{\mathbf{X}}_\mathrm{d}\} = \{\mathbf{0}\}, \tag{7.62}$$

which is a matrix delay differential equation (DDE) with periodic coefficients in matrices $[\mathbf{C}]$, $[\mathbf{K}]$,$[\mathbf{D}]$, and $[\bar{\mathbf{D}}]$. The corresponding elements of these matrices are repeated from Eq. (7.24).

$$C_{jk} = \left(\frac{\partial \phi_j}{\partial \dot{x}_k}\right),\ K_{jk} = \left(\frac{\partial \phi_j}{\partial x_k}\right),\ D_{jk} = \left(\frac{\partial \phi_j}{\partial x_{dk}}\right) \text{and } \bar{D}_{jk} = \left(\frac{\partial \phi_j}{\partial \dot{x}_{dk}}\right). \tag{7.63}$$

It is worth mentioning that all the elements need to be evaluated at the known solution $\{\mathbf{X}\}$. Thus, in general, an $n$th order autonomous delay differential equation can be expressed by $n$ number of 1st order DDEs as follows:

$$\{\dot{\mathbf{x}}(t)\} = [\mathbf{N}(t)]\{\mathbf{x}(t)\} + [\mathbf{W}(t)]\{\mathbf{x}(t-d)\}, \tag{7.64}$$

where $d$ is the time delay. $[\mathbf{N}]$ and $[\mathbf{W}]$ must be time-periodic matrices with a time period $T$, i.e., $[\mathbf{N}(t + T)] = [\mathbf{N}(t)]$ and $[\mathbf{W}(t + T)] = [\mathbf{W}(t)]$.

The first step in the semi-discretization method for DDEs is to divide a single time period $T$ into $k$ number of small intervals, with each interval having length $\Delta t = T/k$ (Figure 7.4a and b). The state $\{\mathbf{x}(t)\}$ is predicted between time $t_i$ and $t_{i+1}$ ($i = 0$, 1, 2 …) using the information from the discretized history of the function between time $t_{i-m}$ and $t_{i-m+1}$. Let us denote $\{\mathbf{x}_j\}$ ($j$ is an integer) as the state of $\{\mathbf{x}(t)\}$ at time $t_j$, i.e., $\{\mathbf{x}_j\} = \{\mathbf{x}(t_j)\}$. The delayed term is approximated as a weighted linear combination of the delayed discrete values $\{\mathbf{x}_{i-m}\}$ and $\{\mathbf{x}_{i-m+1}\}$, where $m$ is a floor function that rounds positive numbers toward zero and is expressed symbolically by

$$m = \left\lfloor \frac{d + \Delta t / 2}{\Delta t} \right\rfloor. \tag{7.65}$$

The integer $m$ can be considered to be an approximation parameter, regarding the time delay. In the $i$th interval, Eq. (7.55) can be approximated to as:

$$\{\dot{\mathbf{x}}(t)\} = [\mathbf{N}_i(t)]\{\mathbf{x}(t)\} + [\mathbf{W}_i(t)]\{\mathbf{x}_{d,i}\}, \tag{7.66}$$

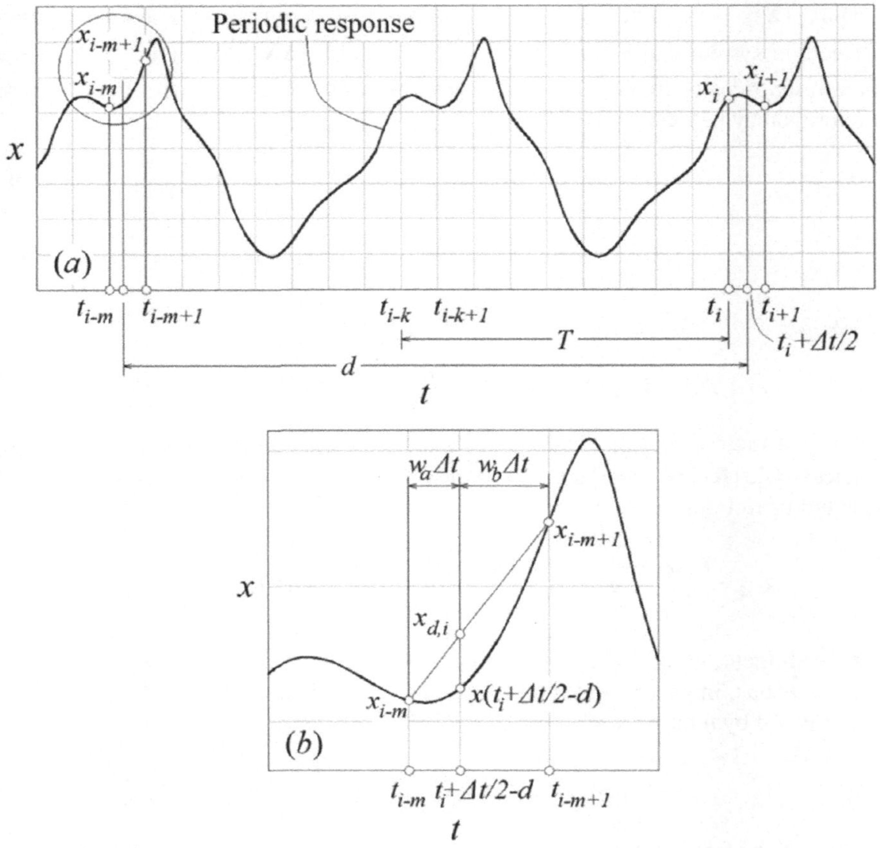


**FIGURE 7.4** (a) Approximation of the time delay term under periodic response. *T* denotes the time period, and *d* denotes the time delay. (b) Zoomed view of the encircled part shown in Figure 7.4a. The stability criteria depend on real and imaginary parts of the eigenvalues of the transition matrix.

$$\text{where } \left[\mathbf{N}_i\right]=\frac{1}{\Delta t}\int_{t_i}^{t_{i+1}}\left[\mathbf{N}(t)\right]dt \text{ and } \left[\mathbf{W}_i\right]=\frac{1}{\Delta t}\int_{t_i}^{t_{i+1}}\left[\mathbf{W}(t)\right]dt. \tag{7.67}$$

The delay term is approximated to as:

$$\{\mathbf{x}(t-d)\}\approx\left\{\mathbf{x}\left(t_i+\Delta t/2-d\right)\right\}\approx w_b\left\{\mathbf{x}_{i-m}\right\}+w_a\left\{\mathbf{x}_{i-m+1}\right\}=\left\{\mathbf{x}_{d,i}\right\} \tag{7.68}$$

Then, the solution of Eq. (7.57) for the initial condition $\{\mathbf{x}(t_i)\}=\{\mathbf{x}_i\}$ can be approximately expressed as:

$$\{\mathbf{x}(t)\} = \left[\exp\left(\mathbf{N}_i(t-t_i)\right)\right]$$
$$\left(\{\mathbf{x}_i\} + \left[\mathbf{N}_i^{-1}\right]\left[\mathbf{W}_i\right]\{\mathbf{x}_{d,i}\}\right) + \left[\mathbf{N}_i^{-1}\right]\left[\mathbf{W}_i\right]\{\mathbf{x}_{d,i}\}. \tag{7.69}$$

The weights of $\{\mathbf{x}_{i-m+1}\}$ and $\{\mathbf{x}_{i-m}\}$ are, respectively, expressed by

$$w_a = \frac{\Delta t(1/2+m)-d}{\Delta t} \text{ and } w_b = \frac{d+\Delta t(1/2-m)}{\Delta t}. \tag{7.70}$$

At time $t = t_{i+1}$, $\{\mathbf{x}(t_{t+1})\} = \{\mathbf{x}_{i+1}\}$. Hence, from Eq. (7.60b),

$$(\mathbf{x}_{i+1}\} = \left[\mathbf{A}_i\right]\{\mathbf{x}_i\} + w_b\left[\mathbf{B}_i\right]\{\mathbf{x}_{i-m}\} + w_a\left[\mathbf{B}_i\right]\{\mathbf{x}_{i-m+1}\}, \tag{7.71}$$

$$\left[\mathbf{A}_i\right] = \exp\left(\left[\mathbf{N}_i\right]\Delta t\right), \tag{7.72}$$

and

$$\left[\mathbf{B}_i\right] = \left(\exp\left(\left[\mathbf{N}_i\right]\Delta t\right) - \mathbf{I}\right)\left[\mathbf{N}_i^{-1}\right]\left[\mathbf{W}_i\right]. \tag{7.73}$$

The connection between the states of the system at time instants $t_i$ and $t_{i+1}$ is presented as a discrete map

$$\{\mathbf{y}_{i+1}\} = \left[\mathbf{C}_i\right]\{\mathbf{y}_i\}, \tag{7.74}$$

$$\text{where } \{\mathbf{y}_{i+1}\} = \begin{bmatrix} \{\mathbf{X}_{i+1}\} \\ \vdots \\ \{\mathbf{X}_i\} \\ \{\mathbf{X}_{i-m+1}\} \end{bmatrix} \text{ and } \{\mathbf{y}_i\} = \begin{bmatrix} \{\mathbf{X}_i\} \\ \vdots \\ \{\mathbf{X}_{i-1}\} \\ \{\mathbf{X}_{i-m}\} \end{bmatrix}. \tag{7.75}$$

The coefficient matrix is expressed by Eq. (7.76) (Insperger and Stepan, 2004) given below.

$$\left[\mathbf{C}_i\right] = \begin{pmatrix} \left[\mathbf{A}_i\right] & 0 & 0 & \cdots & 0 & w_a\left[\mathbf{B}_i\right] & w_b\left[\mathbf{B}_\mathrm{i}\right] \\ [\mathbf{I}] & 0 & 0 & \cdots & 0 & 0 & 0 \\ 0 & [\mathbf{I}] & 0 & \cdots & 0 & 0 & 0 \\ \vdots & \vdots & \vdots & \vdots & \vdots & \vdots & \vdots \\ 0 & 0 & 0 & \cdots & 0 & 0 & 0 \\ 0 & 0 & 0 & \cdots & [\mathbf{I}] & 0 & 0 \\ 0 & 0 & 0 & \cdots & 0 & [\mathbf{I}] & 0 \end{pmatrix} \tag{7.76}$$

The transition matrix $\left([\hat{\Gamma}]\right)$ is calculated over the principal period $T=k\Delta t$.

$$\left(\{\mathbf{y}_k\} = [\hat{\Gamma}]\{\mathbf{y}_0\},\right. \tag{7.77}$$

where $[\hat{\Gamma}]$ is given by the product

$$[\hat{\Gamma}] = \left[\mathbf{C}_{k-1}\right]\left[\mathbf{C}_{k-2}\right]\ldots\left[\mathbf{C}_1\right]\left[\mathbf{C}_0\right] \tag{7.78}$$

Thus, a finite-dimensional approximation of the monodromy operator in the infinite-dimensional version of the Floquet theory (Hale et al., 1990) is achieved. It may be noted that the integer $k$ determines the number of matrices to be multiplied in Eq. (7.78) and $m$ determines the size of these matrices. For systems with a large principal period (i.e., with large $k$) and with a large delay (i.e., with large $m$), the determination of the transition matrix might be time-consuming.

According to Floquet's theory of stability analysis, if the absolute values of all eigenvalues of the transition matrix $\left([\hat{\Gamma}]\right)$ are less than unity, then the periodic solution is stable. If at least one of the eigenvalues has a magnitude greater than one, then the periodic solution is unstable. The way the eigenvalues leave the unit circle determines the nature of the bifurcations (Figure 7.1).

## REFERENCES

Bhattiprolu U, Bajaj A. and Davies K.P. (2016), Periodic response predictions of beams on nonlinear and viscoelastic unilateral foundations using incremental harmonic balance method, *International Journal of Solids and Structures*, Vol. 99, pp. 28–39.

Hale J.K., Verduyn L. and Sjoerd M., (1990), *Introduction to Functional Differential Equations*, Springer.

Hsu C.S., (1972), Impulsive parametric excitation: theory, *Journal of Applied Mechanics*, Vol. 39, pp. 551–558.

Hsu C.S., (1974), On approximating a general impulsive parametric excitation: theory, *Journal of Mathematical Analysis and Applications*, Vol. 45, pp. 234–251.

Hsu C.S. and Cheng W.H., (1973), Applications of the theory of impulsive parametric excitation and new treatments of general parametric excitation problems, *Journal of Applied Mechanics*, Vol. 40, pp. 78–86.

Huai X., Xianren K., Haiqin L. and Zhenguo Y., (2017), Vibration analysis of nonlinear systems with the bilinear hysteretic oscillator by using incremental harmonic balance method, *Communications in Nonlinear Science and Numerical Simulation*, Vol. 42, pp. 437–450.

Insperger T. and Stépán G., (2004), Updated semi-discretization method for periodic delay-differential equations with discrete delay, *International Journal of Numerical Methods in Engineering*, Vol. 61, pp. 117–141.

Ji J.C. and Hansen C.H., (2006), Stability and dynamics of a controlled Van der Pol–Duffing oscillator, *Chaos, Solitons and Fractals*, Vol. 28, pp. 555–570.

Leung A.Y.T. and Chui S.K., (1995), Nonlinear vibration of coupled Duffing oscillators by an improved incremental harmonic balance method, *Journal of Sound and Vibration*, Vol. 181(4), pp. 619–633.

Liu G., Lv Z.R., Liu J.K. and Chen Y.M., (2018), Quasi-periodic aeroelastic response analysis of an airfoil with external store by incremental harmonic balance method, *International Journal of Nonlinear Mechanics*, Vol. 100, pp. 10–19.

Plaut R.H. and Hsieh J.C., (1987a), Chaos in a mechanism with time delays under parametric and external excitation, *Journal of Sound and Vibration*, Vol. 114, pp. 73–90.

Plaut R.H. and Hsieh J.C., (1987b), Nonlinear structural vibrations involving a time delay in damping, *Journal of Sound and Vibration*, Vol. 117, pp. 497–510.

Ranjbarzadeh H. and Kakavand F., (2019), Determination of nonlinear vibration of 2DOF system with an asymmetric piecewise-linear compression spring using incremental harmonic balance method, *European Journal of Mechanics: A/Solids*, Vol. 73, pp. 161–168.

Seydel R. (1988), *From Equilibrium to Chaos, Practical Bifurcation and Stability Analysis*, Elsevier Science Publishing Co., Inc. New York.

Shen Y.J., Wen S.F., Li X.H., Yang S.P. and Xing H.J., (2016), Dynamical analysis of fractional-order nonlinear oscillator by incremental harmonic balance method, *Nonlinear Dynamics*, Vol. 85, pp. 1457–1467.

Tsuda Y., Tamura H., Sueoka A. and Fujii T., (1992), Chaotic behaviour of a nonlinear vibrating system with a retarded argument, *JSME International Journal of Vibration, Control Engineering, Engineering for Industry*, Vol. 35, pp. 259–267.

Wang S., Hua L., Yang C., Han X. and Su Z., (2019), Applications of incremental harmonic balance method combined with equivalent piecewise linearization on vibrations of nonlinear stiffness systems, *Journal of Sound and Vibration*, Vol. 441, pp. 111–125.

Wang Y., Zhang L., and Zhou J., (2020), Incremental harmonic balance method for periodic forced oscillation of a dielectric elastomer balloon, *Applied Mathematics and Mechanics*, Vol. 41, pp. 459–470.

Wei L.S., Wang Y.Z., and Wang Y.S., (2020), Nonreciprocal transmission of nonlinear elastic wave metamaterials by incremental harmonic balance method, *International Journal of Mechanical Sciences*, DOI: 10.1016/j.ijmecsci.2020.105433

Yuanping L. and Siyu C., (2016), Periodic solution and bifurcation of a suspension vibration system by incremental harmonic balance and continuation method, *Nonlinear Dynamics*, Vol. 83, pp 941–950, DOI 10.1007/s11071-015-2378-5

Zhou S., Song G., Ren Z. and Wen B., (2016), Nonlinear analysis of a parametrically excited beam with intermediate support by using multi-dimensional incremental harmonic balance method, *Chaos, Solitons & Fractals*, Vol. 93, pp. 207–222.

# 8 Solution to the Dirac Equation

*S.K. Pandey*

**CONTENTS**



## 8.1 INTRODUCTION

Differential equations have widespread applications in engineering, mathematics, and physics [1–5] and, in this chapter, we study a solution to the Dirac equation.

The Dirac equation is a well-known equation of physics, and, mathematically, it is a partial differential equation.

The Dirac equation [6–8] first appeared in a paper entitled 'The quantum theory of electron,' written by P.A.M. Dirac who was, by training, an electrical engineer.

The Klein-Gordon equation has a second-order derivative with respect to time, but it was unsuccessful in describing the electron. Dirac was motivated to obtain an equation having only a first-order derivative with respect to time, and this idea led him to the factorization of the Klein-Gordon equation. Thus, he obtained a new equation for the electron, which is now known as the Dirac equation.

The Dirac equation is a very mysterious equation of quantum physics. Many authors, including Dirac himself, have studied this equation, but the underlying mysteries of this equation have not been revealed so far.

Motivated by an earlier work of de Broglie, Toyoki Koga [9–12] studied the Dirac equation and provided a deterministic solution. In the past we have studied Koga's solution using the technique of geometric algebra [13–15]. But the purpose of this chapter is different.

The unique feature of this work is that here we use Koga's technique to solve the massless Dirac equation in four dimensions without using the technique of geometric algebra. The solution to the massless Dirac equation given in this work should be considered as a mere clue to the existence of a massless spinning Dirac field with imaginary energy. Hopefully, in the near future, somebody will come up with an empirical justification for the existence of such a field. The work presented here may or may not fit in the conventional realm of modern physics but its importance lies in the fact that it provides a clue to the existence of a massless spinning Dirac field with imaginary energy.

In Section 8.2, we give a brief introduction to Koga's solution to the Klein-Gordon equation, which is used to solve the Dirac equation by Koga's technique, and, in Sections 8.3 and 8.4, we find solutions to the massless and anti-massless Dirac field in four dimensions.

In Section 8.5, we provide the results, and in Section 8.6 we discuss the results obtained in this chapter. In the last section (Section 8.7), we provide the conclusions.

## 8.2 PRELIMINARIES

The Klein-Gordon equation for a free particle has the following form [10]:

$$\left(\hbar^2 \frac{\partial^2}{\partial t^2} - \hbar^2 c^2 \left(\frac{\partial^2}{\partial x^2} + \frac{\partial^2}{\partial y^2} + \frac{\partial^2}{\partial z^2}\right) + m^2 c^4\right)\phi = 0$$

Here we work in Minkowski space. A point is given by

$$(ct, x, y, z) = (x^0, x^1, x^2, x^3).$$

The Minkowski metric is given by

$$\eta_{ij} dx^i dx^j = (dx^0)^2 - (dx^1)^2 - (dx^2)^2 - (dx^3)^2$$

In Koga's notation [9], the Dirac equation for a free electron can be given as

$$\left[i\hbar c\beta \frac{\partial}{\partial t} + i\hbar c\beta\left(\alpha_x \frac{\partial}{\partial x} + \alpha_y \frac{\partial}{\partial y} + \alpha_z \frac{\partial}{\partial z}\right) + mc^2\right]\Psi = 0$$

where

$$\alpha_x^2 = \alpha_x^2 = \alpha_x^2 = \beta^2 = I_4$$

$$\alpha_x = \rho\sigma_x,\ \alpha_y = \rho\sigma_y,\ \alpha_z = \rho\sigma_z$$

$\alpha_x\alpha_y + \alpha_y\alpha_x = 0$, $\alpha_x\beta + \beta\alpha_x = 0$, and so on.

$$\Psi = \begin{pmatrix} \Psi_1 \\ \Psi_2 \\ \Psi_3 \\ \Psi_4 \end{pmatrix}, \ I_4 = \begin{pmatrix} 1 & 0 & 0 & 0 \\ 0 & 1 & 0 & 0 \\ 0 & 0 & 1 & 0 \\ 0 & 0 & 0 & 1 \end{pmatrix}, \ \beta = \begin{pmatrix} 1 & 0 & 0 & 0 \\ 0 & 1 & 0 & 0 \\ 0 & 0 & -1 & 0 \\ 0 & 0 & 0 & -1 \end{pmatrix}$$

$$\sigma_x = \begin{pmatrix} 0 & 1 & 0 & 0 \\ 1 & 0 & 0 & 0 \\ 0 & 0 & 0 & 1 \\ 0 & 0 & 1 & 0 \end{pmatrix}, \ \sigma_y = \begin{pmatrix} 0 & -i & 0 & 0 \\ i & 0 & 0 & 0 \\ 0 & 0 & 0 & -i \\ 0 & 0 & i & 0 \end{pmatrix}$$

$$\sigma_z = \begin{pmatrix} 1 & 0 & 0 & 0 \\ 0 & -1 & 0 & 0 \\ 0 & 0 & 1 & 0 \\ 0 & 0 & 0 & -1 \end{pmatrix}, \rho = \begin{pmatrix} 0 & 0 & 1 & 0 \\ 0 & 0 & 0 & 1 \\ 1 & 0 & 0 & 0 \\ 0 & 1 & 0 & 0 \end{pmatrix}$$

Koga [9–11] has given a solution to the Klein-Gordon equation, which for a free particle at rest centered on the origin of the rest frame can be given as

$$\phi = a \exp\left(\frac{iS}{\hbar}\right)$$

where $S = -Ect$, $E^2 = m^2c^2 - \hbar^2\kappa^2$. Here $Ec$ is the energy of the field,

$$\text{and } a = \frac{\exp\left(-\kappa\left|r\right|\right)}{\left|r\right|}, \ \kappa > 0.$$

$$r = \left|\vec{r}\right| \text{ and } \vec{r} = x\hat{i} + y\hat{j} + z\hat{k}.$$

It should be noted that Koga's technique [9–11] is to produce a solution to the Dirac equation, using this solution of the Klein-Gordon equation.

$$\text{Let } \Psi = \left[i\hbar c\beta\frac{\partial}{\partial t} + i\hbar c\beta\left(\alpha_x\frac{\partial}{\partial x} + \alpha_y\frac{\partial}{\partial y} + \alpha_z\frac{\partial}{\partial z}\right) - mc^2\right]\varphi$$

Here, $\varphi = a\exp\left(\frac{iS}{\hbar}\right)A_j\exp\left(i\theta_j\right)$ ($j$ = 1, 2, 3, 4). Then $\Psi$ is a solution of the Dirac equation. It may be noted that each component of $\phi$ is a solution to the Klein-Gordon equation.

The Dirac matrices have several representations. However, Koga worked with the well-known standard representation [16, 17]. It is seen that

$$\beta\alpha_x = \begin{pmatrix} 0 & 0 & 0 & -1 \\ 0 & 0 & 1 & 0 \\ 0 & 1 & 0 & 0 \\ 1 & 0 & 0 & 0 \end{pmatrix}, \ \beta\alpha_y = \begin{pmatrix} 0 & 0 & 0 & i \\ 0 & 0 & -i & 0 \\ 0 & -i & 0 & 0 \\ i & 0 & 0 & 0 \end{pmatrix}, \ \beta\alpha_z = \begin{pmatrix} 0 & 0 & -1 & 0 \\ 0 & 0 & 0 & 1 \\ 1 & 0 & 0 & 0 \\ 0 & -1 & 0 & 0 \end{pmatrix}.$$

If we take

$$\beta = \gamma^0, \ \beta\alpha_x = \gamma^1, \ \beta\alpha_y = \gamma^2 \text{ and } \beta\alpha_z = \gamma^3$$

then we get the well-known standard representation of the Dirac matrices.

It may be noted that here $\Psi = \begin{pmatrix} \Psi_1 \\ \Psi_2 \\ \Psi_3 \\ \Psi_4 \end{pmatrix}$ is a four-component spinor. However, in the case of the massless field, $\Psi$ reduces to $\psi = \begin{pmatrix} \psi_1 \\ \psi_2 \end{pmatrix}$ which is a two- component spinor.

It may be worth mentioning that recently we have studied canonical representations of Koga's solution to the Dirac equation [18]. The work presented in [18] is based on our earlier work [13–15].

Now, we shall use this technique in the next section to obtain a solution to the massless Dirac equation.

## 8.3 SOLUTION TO THE MASSLESS FIELD

The Klein-Gordon equation for a massless field takes the following form:

$$\left(\hbar^2 \frac{\partial^2}{\partial t^2} - \hbar^2 c^2 \left( \frac{\partial^2}{\partial x^2} + \frac{\partial^2}{\partial y^2} + \frac{\partial^2}{\partial z^2} \right)\right)\phi = 0 \tag{8.1}$$

Following Koga (8.1), this can be written as

$$D_0 \, D_1 \phi = 0 = D_1 \, D_0 \phi$$

where $D_1 = \left( \hbar \frac{\partial}{\partial t} - \hbar c \hat{\sigma}_1 \frac{\partial}{\partial x} - \hbar c \hat{\sigma}_2 \frac{\partial}{\partial y} - \hbar c \hat{\sigma}_3 \frac{\partial}{\partial z} \right)$

and $D_0 = \left( \hbar \frac{\partial}{\partial t} + \hbar c \hat{\sigma}_1 \frac{\partial}{\partial x} + \hbar c \hat{\sigma}_2 \frac{\partial}{\partial y} + \hbar c \hat{\sigma}_3 \frac{\partial}{\partial z} \right)$

Here,

$$\hat{\sigma}_1 = \begin{pmatrix} 0 & 1 \\ 1 & 0 \end{pmatrix}, \ \hat{\sigma}_2 = \begin{pmatrix} 0 & -i \\ i & 0 \end{pmatrix}, \ \hat{\sigma}_3 = \begin{pmatrix} 1 & 0 \\ 0 & -1 \end{pmatrix}$$

are known as the Pauli matrices.

It may be noted that both $D_1\psi = 0$ and $D_0\psi = 0$ give the legitimate massless Dirac equation; however, we shall take $D_0\psi = 0$ as the massless Dirac equation. We shall consider $D_1\psi = 0$ in the next section.

Now, if $\varphi$ is a solution of the Klein-Gordon equation and $\psi = D_1\phi$, then $\psi$ satisfies $D_0\psi = 0$, which is the massless Dirac equation in four dimensions.

In the limit considered in this chapter, we have $\phi = a\exp(i\kappa c t)$ and the energy of the field does not vanish even at this limit, so we assume that this approximation is mathematically valid.

In the conventional realm of modern physics, the energy of a massless particle at rest becomes zero and such a particle has actually no meaning, and therefore a rest frame does not exist and such a particle is always assumed to travel with the speed of light in all frames.

It is worth mentioning that, for $\hbar^2 = 0$, the expression for energy given in Section 8.2 reduces to the usual expression for energy, which is used for point particles in the conventional theory of special relativity [9]. This can be seen below. We have

$$E^2 = \frac{m^2c^2}{1-\frac{u^2}{c^2}} \quad \text{(taking } \hbar^2 = 0\text{)}.$$

This implies that $E^2c^2 - E^2u^2 = m^2c^4$. This gives $E^2c^2 = p^2c^2 + m^2c^4$ (as already mentioned here, $cE$ is the energy). Therefore, in conventional theory, if a massless particle is considered at rest then its energy vanishes; hence, it can never be at rest. However, in Koga's theory the situation is different.

Clearly in this theory the energy of the massless field is imaginary, and the energy of the massless field does not vanish even if we assume that it is at rest. The energy remains imaginary. Clearly the energy of a massless field at rest becomes $\pm i\hbar\kappa c$.

Now $\psi = D_1\phi$ gives

$$\psi = \left(\hbar\frac{\partial}{\partial t} - \hbar c\sigma_1\frac{\partial}{\partial x} - \hbar c\sigma_2\frac{\partial}{\partial y} - \hbar c\sigma_3\frac{\partial}{\partial z}\right)\phi$$

$$= \left(\hbar\frac{\partial}{\partial t} - \hbar c\hat{\sigma}_1\frac{\partial}{\partial x} - \hbar c\hat{\sigma}_2\frac{\partial}{\partial y} - \hbar c\hat{\sigma}_3\frac{\partial}{\partial z}\right)a\exp\left(\frac{iS}{\hbar}\right)A_j\exp(i\theta_j)$$

or

$$\psi = \left[-iEc - \hbar c\hat{\sigma}_1 R_x - \hbar c\hat{\sigma}_2 R_y - \hbar c\hat{\sigma}_3 R_z\right]a\exp\left(\frac{iS}{\hbar}\right)A_j\exp(i\theta_j) \qquad (8.2)$$

Here, $R_x = x\left(\frac{1}{r^2} + \frac{\kappa}{r}\right)$, $R_y = y\left(\frac{1}{r^2} + \frac{\kappa}{r}\right)$, $R_z = z\left(\frac{1}{r^2} + \frac{\kappa}{r}\right)$

It may be noted that $\psi$ given by Eq. (8.2) is a two-component spinor, which is commonly known as a weyl spinor [8]. Therefore, we have $\psi = \begin{pmatrix} \psi_1 \\ \psi_2 \end{pmatrix}$ and $A_j \exp(i\theta_j) = \begin{pmatrix} A_1 \exp(i\theta_1) \\ A_2 \exp(i\theta_2) \end{pmatrix}$. As mentioned by Koga, the theory does not determine the real constants, $A_j$ and $\theta_j$. So, in order to get a physically meaningful solution, we have to choose proper values for these constants.

Now we shall find two solutions for the Dirac equation corresponding to the following two values of $\begin{pmatrix} A_1 \exp(i\theta_1) \\ A_2 \exp(i\theta_2) \end{pmatrix} = \begin{pmatrix} 1 \\ 0 \end{pmatrix}$ and $\begin{pmatrix} A_1 \exp(i\theta_1) \\ A_2 \exp(i\theta_2) \end{pmatrix} = \begin{pmatrix} 0 \\ 1 \end{pmatrix}$. It may be noted that $\begin{pmatrix} 1 \\ 0 \end{pmatrix}$ and $\begin{pmatrix} 0 \\ 1 \end{pmatrix}$ form a basis set for the two-spinor space and these are used in conventional works on the Dirac equation [7, 8] dealing with two-component spinors.

**Case I:** Taking $\begin{pmatrix} A_1 \exp(i\theta_1) \\ A_2 \exp(i\theta_2) \end{pmatrix} = \begin{pmatrix} 1 \\ 0 \end{pmatrix}$, Eq. (8.1) becomes (we have $Ec = i\kappa\hbar c$)

$$\begin{pmatrix} \psi_1 \\ \psi_2 \end{pmatrix} = a\exp(\kappa ct)\left[\kappa\hbar c\begin{pmatrix} 1 & 0 \\ 0 & 1 \end{pmatrix} - \hbar c\begin{pmatrix} 0 & 1 \\ 1 & 0 \end{pmatrix}R_x - \hbar c\begin{pmatrix} 0 & -i \\ i & 0 \end{pmatrix}R_y - \hbar c\begin{pmatrix} 1 & 0 \\ 0 & -1 \end{pmatrix}R_z\right]\begin{pmatrix} 1 \\ 0 \end{pmatrix}$$

One may easily note that $\begin{pmatrix} 1 & 0 \\ 0 & 1 \end{pmatrix}\begin{pmatrix} 1 \\ 0 \end{pmatrix} = \begin{pmatrix} 1 \\ 0 \end{pmatrix}, \begin{pmatrix} 0 & 1 \\ 1 & 0 \end{pmatrix}\begin{pmatrix} 1 \\ 0 \end{pmatrix} = \begin{pmatrix} 0 \\ 1 \end{pmatrix}$

$$\begin{pmatrix} 0 & -i \\ i & 0 \end{pmatrix}\begin{pmatrix} 1 \\ 0 \end{pmatrix} = \begin{pmatrix} 0 \\ i \end{pmatrix} \text{ and } \begin{pmatrix} 1 & 0 \\ 0 & -1 \end{pmatrix}\begin{pmatrix} 1 \\ 0 \end{pmatrix} = \begin{pmatrix} 1 \\ 0 \end{pmatrix}$$

Therefore

$$\begin{pmatrix} \psi_1 \\ \psi_2 \end{pmatrix} = a\exp(\kappa ct)\left[\kappa\hbar c\begin{pmatrix} 1 \\ 0 \end{pmatrix} - \hbar c\begin{pmatrix} 0 \\ 1 \end{pmatrix}R_x - \hbar c\begin{pmatrix} 0 \\ i \end{pmatrix}R_y - \hbar c\begin{pmatrix} 1 \\ 0 \end{pmatrix}R_z\right]$$

$$= a\exp(\kappa ct)\begin{bmatrix} \kappa\hbar c - \hbar cR_z \\ -\hbar cR_x - i\hbar cR_y \end{bmatrix}$$

Now we have

$$\psi_1 = a\exp(\kappa ct)(\kappa\hbar c - \hbar cR_z)$$

and $\psi_2 = a\exp(\kappa ct)(-\hbar cR_x - i\hbar cR_y)$

**Case II:** Taking $\begin{pmatrix} A_1 \exp(i\theta_1) \\ A_2 \exp(i\theta_2) \end{pmatrix} = \begin{pmatrix} 0 \\ 1 \end{pmatrix}$, Eq. (8.1) becomes

$$\begin{pmatrix} \psi_1 \\ \psi_2 \end{pmatrix} = a\exp(\kappa c t)\left[\kappa\hbar c\begin{pmatrix} 1 & 0 \\ 0 & 1 \end{pmatrix} - \hbar c\begin{pmatrix} 0 & 1 \\ 1 & 0 \end{pmatrix}R_x - \hbar c\begin{pmatrix} 0 & -i \\ i & 0 \end{pmatrix}R_y - \hbar c\begin{pmatrix} 1 & 0 \\ 0 & -1 \end{pmatrix}R_z\right]\begin{pmatrix} 0 \\ 1 \end{pmatrix}$$

One may easily note that $\begin{pmatrix} 1 & 0 \\ 0 & 1 \end{pmatrix}\begin{pmatrix} 0 \\ 1 \end{pmatrix} = \begin{pmatrix} 0 \\ 1 \end{pmatrix}$, $\begin{pmatrix} 0 & 1 \\ 1 & 0 \end{pmatrix}\begin{pmatrix} 0 \\ 1 \end{pmatrix} = \begin{pmatrix} 1 \\ 0 \end{pmatrix}$

$$\begin{pmatrix} 0 & -i \\ i & 0 \end{pmatrix}\begin{pmatrix} 0 \\ 1 \end{pmatrix} = \begin{pmatrix} -i \\ 0 \end{pmatrix} \text{ and } \begin{pmatrix} 1 & 0 \\ 0 & -1 \end{pmatrix}\begin{pmatrix} 0 \\ 1 \end{pmatrix} = \begin{pmatrix} 0 \\ -1 \end{pmatrix}$$

Therefore

$$\begin{aligned} \begin{pmatrix} \psi_1 \\ \psi_2 \end{pmatrix} &= a\exp(\kappa c t)\left[\kappa\hbar c\begin{pmatrix} 0 \\ 1 \end{pmatrix} - \hbar c\begin{pmatrix} 1 \\ 0 \end{pmatrix}R_x - \hbar c\begin{pmatrix} -i \\ 0 \end{pmatrix}R_y - \hbar c\begin{pmatrix} 0 \\ -1 \end{pmatrix}R_z\right] \\ &= a\exp(\kappa c t)\begin{bmatrix} -\hbar c R_x + i\hbar c R_y \\ \kappa\hbar c - \hbar c R_z \end{bmatrix} \end{aligned}$$

Now we have

$$\psi_1 = a\exp(\kappa c t)(-\hbar c R_x + i\hbar c R_y)$$

and $\psi_2 = a\exp(\kappa c t)(\kappa\hbar c - \hbar c R_z)$

## 8.4 SOLUTION TO THE ANTI-MASSLESS FIELD

The Dirac equation for the anti-massless field can be given as

$$\left(\hbar\frac{\partial}{\partial t} - \hbar c\hat{\sigma}_1\frac{\partial}{\partial x} - \hbar c\hat{\sigma}_2\frac{\partial}{\partial y} - \hbar c\hat{\sigma}_3\frac{\partial}{\partial z}\right)\psi = 0$$

In this case, following the same method as described above, the solution is given by

$$\psi = D_0\phi$$

or $\psi = \left(\hbar\frac{\partial}{\partial t} + \hbar c\sigma_1\frac{\partial}{\partial x} + \hbar c\sigma_2\frac{\partial}{\partial y} + \hbar c\sigma_3\frac{\partial}{\partial z}\right)\phi$

$$= \left( \hbar \frac{\partial}{\partial t} + \hbar c \hat{\sigma}_1 \frac{\partial}{\partial x} + \hbar c \hat{\sigma}_2 \frac{\partial}{\partial y} + \hbar c \hat{\sigma}_3 \frac{\partial}{\partial z} \right) a \exp\left( \frac{iS}{\hbar} \right) A_j \exp(i\theta_j)$$

or

$$\psi = \left[ -iEc + \hbar c \hat{\sigma}_1 R_x + \hbar c \hat{\sigma}_2 R_y + \hbar c \hat{\sigma}_3 R_z \right] a \exp\left( \frac{iS}{\hbar} \right) A_j \exp(i\theta_j) \tag{8.3}$$

Taking $\begin{pmatrix} A_1 \exp(i\theta_1) \\ A_2 \exp(i\theta_2) \end{pmatrix} = \begin{pmatrix} 1 \\ 0 \end{pmatrix}$, we get

$$\begin{pmatrix} \psi_1 \\ \psi_2 \end{pmatrix} = a \exp(\kappa c t) \left[ -\kappa \hbar c \begin{pmatrix} 1 & 0 \\ 0 & 1 \end{pmatrix} + \hbar c \begin{pmatrix} 0 & 1 \\ 1 & 0 \end{pmatrix} R_x + \hbar c \begin{pmatrix} 0 & -i \\ i & 0 \end{pmatrix} R_y + \hbar c \begin{pmatrix} 1 & 0 \\ 0 & -1 \end{pmatrix} R_z \right] \begin{pmatrix} 1 \\ 0 \end{pmatrix}$$

$$= a \exp(\kappa c t) \left[ -\kappa \hbar c \begin{pmatrix} 1 \\ 0 \end{pmatrix} + \hbar c \begin{pmatrix} 0 \\ 1 \end{pmatrix} R_x + \hbar c \begin{pmatrix} 0 \\ i \end{pmatrix} R_y + \hbar c \begin{pmatrix} 1 \\ 0 \end{pmatrix} R_z \right]$$

$$= a \exp(\kappa c t) \begin{bmatrix} -\kappa \hbar c + \hbar c R_z \\ \hbar c R_x + i \hbar c R_y \end{bmatrix}$$

Now we have

$$\psi_1 = a \exp(\kappa c t)(-\kappa \hbar c + \hbar c R_z)$$

and $\psi_2 = a \exp(\kappa c t)(\hbar c R_x + i \hbar c R_y)$

Similarly, taking $\begin{pmatrix} A_1 \exp(i\theta_1) \\ A_2 \exp(i\theta_2) \end{pmatrix} = \begin{pmatrix} 0 \\ 1 \end{pmatrix}$, we get

$$\begin{pmatrix} \psi_1 \\ \psi_2 \end{pmatrix} = a \exp(\kappa c t) \left[ -\kappa \hbar c \begin{pmatrix} 1 & 0 \\ 0 & 1 \end{pmatrix} + \hbar c \begin{pmatrix} 0 & 1 \\ 1 & 0 \end{pmatrix} R_x + \hbar c \begin{pmatrix} 0 & -i \\ i & 0 \end{pmatrix} R_y + \hbar c \begin{pmatrix} 1 & 0 \\ 0 & -1 \end{pmatrix} R_z \right] \begin{pmatrix} 0 \\ 1 \end{pmatrix}$$

$$= a \exp(\kappa c t) \left[ -\kappa \hbar c \begin{pmatrix} 0 \\ 1 \end{pmatrix} + \hbar c \begin{pmatrix} 1 \\ 0 \end{pmatrix} R_x + \hbar c \begin{pmatrix} -i \\ 0 \end{pmatrix} R_y + \hbar c \begin{pmatrix} 0 \\ -1 \end{pmatrix} R_z \right]$$

$$= a \exp(\kappa c t) \begin{bmatrix} \hbar c R_x - i \hbar c R_y \\ -\kappa \hbar c - \hbar c R_z \end{bmatrix}$$

Now we have

$$\psi_1 = a \exp(\kappa c t)(\hbar c R_x - i \hbar c R_y)$$

and $\psi_2 = a \exp(\kappa c t)(-\kappa \hbar c - \hbar c R_z)$

## 8.5 RESULTS

We have obtained the following solutions to the massless field:

a) $\psi_1 = a\exp(\kappa ct)(\kappa\hbar c - \hbar cR_z)$

$\psi_2 = a\exp(\kappa ct)(-\hbar cR_x - i\hbar cR_y)$

b) $\psi_1 = a\exp(\kappa ct)(-\hbar cR_x + i\hbar cR_y)$

$\psi_2 = a\exp(\kappa ct)(\kappa\hbar c - \hbar cR_z)$

Similarly we obtained the following solutions to the anti-massless field:

c) $\psi_1 = a\exp(\kappa ct)(-\kappa\hbar c + \hbar cR_z)$

$\psi_2 = a\exp(\kappa ct)(\hbar cR_x + i\hbar cR_y)$

d) $\psi_1 = a\exp(\kappa ct)(\hbar cR_x - i\hbar cR_y)$

$\psi_2 = a\exp(\kappa ct)(-\kappa\hbar c - \hbar cR_z).$

## 8.6 DISCUSSION

One may note that the fields given by $\psi_1$ and $\psi_2$ in both cases, namely massless fields and anti-massless fields, are localized fields, and $\psi_1 \to 0$ as $r \to \infty$. Similarly, $\psi_2 \to 0$ as $r \to \infty$.

It may be noted that $a\exp(\kappa ct)$ is nothing but a spherically symmetric scalar field.

Now if we consider a rotation about the $z$-axis, then $\psi_1$ given by (a) and $\psi_2$ given by (b) have rotational symmetry about the $z$-axis. Similarly, $\psi_1$ given by (c) and $\psi_2$ given by (d) have rotational symmetry about the $z$-axis.

Finally, let us consider $\psi_2$ given by (a). Then we can write

$$\psi_2 = e^{i\pi}\left[a\exp(\kappa ct)(\hbar cR_x + i\hbar cR_y)\right]$$

This represents a rotating field through an angle $\theta = \pi$ about the $z$-axis. Since $\psi_1$ given by (b) is just a complex conjugate of $\psi_2$ given by (a), it therefore represents a rotating field with the same angle, although the orientation of rotation is changed.

Similarly, one can see that $\psi_2$ given by (c) represents a rotating field through an angle $\theta = \pi$ about the $z$-axis. As $\psi_1$ given by (d) is just a complex conjugate of $\psi_2$ given by (c), it therefore represents a rotating field with the same angle, although the direction of rotation is reversed.

## 8.7 CONCLUSIONS

We have obtained localized solutions to the massless field and the anti-massless field. The work presented in this chapter may or may not be applicable to the conventional realm of modern physics; however, it gives a clue as to the existence of a massless spinning Dirac field with imaginary energy.

## 8.8 ACKNOWLEDGMENTS

This author is highly thankful to A. Pandit for his support.

## REFERENCES

1. Farlow, S.J. (1993). *Partial Differential Equations for Scientists and Engineers*, Dover Publications Inc.
2. Goodwine, B. (2011). *Engineering Differential Equations Theory and Applications*, Springer.
3. Kalbaugh, D.V. (2017). *Differential Equations for Engineers: The Essentials*, CRC Press.
4. Struthers, A. and Potter, M. (2019). *Differential Equations for Scientists and Engineers*, Springer.
5. Betounes (2010). *Differential Equations: Theory and Applications*, Springer.
6. Dirac, P.A.M. (1981). *The Principles of Quantum Mechanics*, Carendon Press.
7. Thaller, B. (1992). *The Dirac Equation*, Springer.
8. Bulbul, B., Sezer, M., and Greiner, W (1990). *Relativistic Quantum Mechanics Wave Equations*, Springer.
9. Koga, T. (1981). *Foundations of Quantum Physics*, Woods and Jones.
10. Koga, T. (1983). *Inquiries into Foundations of Quantum Physics*, Woods and Jones.
11. Koga, T. (1975). A rational interpretation of the dirac equation for the electron, *Int. J. Theo. Phys.*, 13 (4), 271–278.
12. Koga, T. (1975). Representation of spin in the interpretation of the Dirac equation for the electron, *Int. J. Theo. Phys.*, 12, 205–215.
13. Pandey, S.K., and Chakravarti, R.S. (2009). The Dirac equation: an approach through geometric algebra, *Annales de la Fondation Louis de Broglie*, 34 (2), 223–228.
14. Pandey, S.K., and Chakravarti, R.S. (2011). The Dirac equation through geometric algebra: some implications, Annales de la Fondation Louis de Broglie, 36, 73–77.
15. Pandey, S.K. (2014). Electronic spin: abstract mathematical or real physical phenomenon, arXiv: 1208.5764 [quant-ph].
16. Bjorken, J. D., and Drell, S. D. (1964). *Relativistic Quantum Mechanics*, McGraw Hill Education.
17. Itzykson, C., and Zuber, J.B. (2013). *Quantum Field Theory*, McGraw Hill Education.
18. Pandey, S.K. (2021). Canonical representation of Koga's solution to the Dirac equation. In *Advances and Applications in Computational Mathematics*, ed. Kumar, A., Ram, M., and Srivastva, H.M., (to appear), River Publishers.

# 9 Periodic Solution of a Nonlinear Economic Cycle Model with a Generic Investment Function

*Jun Zhao*

**CONTENTS**



## 9.1 INTRODUCTION

Nonlinear phenomena are widespread in the field of economics and they are generally modeled by some dynamic economic model (Serrano et al. 2012; Riad et al. 2016; Guerrazzi and Sodini 2018; Matsumoto et al. 2018). There are a number of dynamic economic models, which consider the effects of economic fluctuations in either a deterministic sense or a stochastic sense (Amman 1996; Liu et al. 2015; Bashkirtseva et al. 2016, 2018b). Among these models, economic cycle models have been widely investigated with respect to their linear and nonlinear behavior. It is well known that Neoclassical theory and Keynesian theory can apply to economic cycles. In the former, linear economic cycle models are widely investigated to study the stability of economic conditions. By contrast, in the framework of the Keynesian endogenous cycle theory, nonlinear economic cycle models are mainly used to study economic fluctuations. Goodwin (1951) developed a nonlinear model formulation for an economic cycle of consumption and investment. This model was used to study the impact of secular evolution and lags on the economy. Puu and Sushko (2004) studied the expression of an investment function and they proposed a linear-plus-cubic function to describe investment. They employed the nonlinear investment function to include the effects of some governments, e.g., counter-cyclical infrastructure investment to resist depression. Later, Chian et al. (2005) studied a Van der Pol-type economic cycle model, driven by a harmonic exogenous force. The harmonic exogenous force can model external impacts, such as the effects of different-level economies, or the impact of climate changes. Zhao and Li (2016) studied Goodwin's consumption function, considered the form of Puu's investment function, and established a nonlinear discrete dynamical system. The bifurcation and chaos of the dynamical system were investigated. Hattaf et al. (2017) applied a general investment function and developed a delayed business cycle model. The time delays were introduced into the gross product and capital stock. The stability of the economic equilibrium was investigated, as well as the existence of a Hopf bifurcation.

Furthermore, many experts have devoted their efforts to investigating the nonlinear response evolution of some economic cycles in a random environment because some factors in a dynamic economic cycle model are random in nature. Based on the previous work (Goodwin 1951; Puu and Sushko 2004), Li et al. (2007) developed an economic cycle model with uncertainty. They studied the stochastic stability and bifurcation of their proposed model in the case of white noise acting on income. Li and Li (2015) employed a stochastic averaged method to study the nonlinear response evolution of an economic cycle subjected to Poisson impulses. The difference between Poisson impulses and Gaussian white noise is that the former is a continuous stochastic process, whereas the latter represents a discrete stochastic process. The authors proposed that newly issued economic or monetary policies were commonly discrete events. In addition, political or natural events with an economic impact occurred unexpectedly. It is more reasonable that these events can be simulated by discrete Poisson impulses. Tran et al. (2016) generalized the Goodwin-Lotka-Volterra models into a stochastic case. They observed that noise led to a large energy located on the trajectories. Meanwhile, the period of economic cycles became longer. Lin et al. (2016) considered the case of fractional derivatives in an economic cycle model, and the random excitation was narrow band. They used the method of multiple scales to study the effect of economic policy on economic fluctuation, as well as the effect on stationary probability density. Bashkirtseva et al. (2018a) studied a Kaldor-type economic cycle model. The additive and internally parametric noises, related to economic variables, were considered in the model. Some stochastic bifurcations were detected in their study.

On the other hand, a number of harmonic balance methods have recently been proposed for the study of periodic motions of various types of nonlinear dynamical systems. For example, Detroux et al. (2015) developed a harmonic balance method for analyzing the bifurcation of large-scale mechanical systems. The Floquet exponents were used along with bordering techniques. Grolet and Thouverez (2015) applied the Groebner basis to improve the method of harmonic balance for computing solutions to all the polynomial systems. The effect of the vibration of mechanical systems on geometric nonlinearities was investigated. Xu et al. (2017) prescribed computational accuracy and generalized the harmonic balance method to study the periodic motions of one-dimensional nonlinear systems. The frequency-amplitude response of periodic motions was analyzed. The different order harmonic terms in the Fourier series were considered. The harmonic amplitudes were determined by their corresponding quantity. Foupouapouognigni et al. (2017) considered fractional inductance in a nonlinear electromechanical energy harvester. They adopted a harmonic balance method to study its analytical response. The analytical results were validated by the numerical simulation. A parametric study was further studied to make the best choice of system parameters for the improvement of power output. Dai et al. (2018) applied a de-aliasing scheme to the harmonic balance method to obtain the semi-analytical periodic solutions of an aeroelastic system. Yao and Marques (2018) proposed an aeroelastic-harmonic balance method for solving the relevant problems of nonlinear fluid–structure interactions. Limit-cycle oscillations and vortex-induced vibrations were studied in the method proposed by Yao and Marques (2018).

From the above descriptions, nonlinear economic cycle models are powerful tools with which to study economic phenomena, and cases of the models under random disturbance have been widely investigated. However, some factors with a harmonic fluctuation in these nonlinear economic cycles have been considered only rarely. In fact, some regulations from governments on an economic cycle model may have a periodic characteristic, in which a periodic solution should be further studied. This motivates the present study to describe an implicit procedure of the harmonic balance method looking for periodic solutions to nonlinear economic cycle models under harmonic spontaneous adjustment. In the first step, based on the existing models (Goodwin 1951; Puu and Sushko 2004), a nonlinear economic cycle model is developed. The model consists of a linear-quadratic-cubic function on the difference between pre-income and prior income. This nonlinear function is used to describe the generic effect of incomes on investment. Then, the extended implicit solution procedure expresses the periodic solution, using Fourier series. They include complete cosines and sines truncated up to adequate terms. An indexing procedure is developed to establish nonlinear algebraic equations, including the amplitudes of cosines and sines and the frequency and amplitude of harmonic spontaneous adjustments. The solutions to these resulting equations can be obtained with available nonlinear equation solvers in mathematical libraries. Compared with the simulation, the extended implicit procedure can present a close agreement with the simulated periodic solution. A parametric study further shows that the presence of the nonlinear terms leads to the occurrence of different harmonic dynamic responses. When the frequency of harmonic spontaneous adjustment is near the resonance region, the nonlinear economic cycle model has an almost symmetric vibratory mode. However, when the frequency is far from the resonance region, the vibratory response has a non-zero mean, which implies that a constant exists in the periodic solution when the economic cycle evolves.

## 9.2 ECONOMIC CYCLE MODEL

Sushko et al. (2006) advanced the work of Goodwin (1951) on formulating economic cycle models. They proposed a linear-quadratic-cubic function about the difference between pre-income and prior income for investment. This form breaks the symmetry of the economic cycle model and it can simulate more general cases. The expression of the investment function is given below:

$$I_t = I_{0t} + \nu((Y_{t-1} - Y_{t-2}) + \mu(Y_{t-1} - Y_{t-2})^2 - (Y_{t-1} - Y_{t-2})^3) \tag{9.1}$$

where $I_t$ denotes the current investment, $I_{0t}$ denotes the spontaneous investment, $Y_t$ is current income, $Y_{t-1}$ is pre-income, and $Y_{t-2}$ is prior income. $\nu$ denotes the capital output ratio ($\nu > 0$) and $\mu$ is an adjustment coefficient to shape the investment function. The nonlinear terms in Eq. (9.1) can model the countercyclical investment from the government. The government regulates investment on public infrastructure to avoid economic depression. On the other hand, the nonlinear terms can also simulate the benefit from lower investment through a period of great decline. Furthermore, the quadratic term also breaks the symmetry of the economic cycle model, which can simulate a more generic case.

In a similar manner, the consumption function can be expressed using pre-income and prior income:

$$C_t = C_{0t} + (1-s)Y_{t-1} + \varepsilon s Y_{t-2} \tag{9.2}$$

where $C_t$ denotes the current consumption, $C_{0t}$ denotes the spontaneous consumption, $\varepsilon$ denotes the prior saving rate ($0 \le \varepsilon \le 1$), and $s$ denotes the added saving rate ($0 \le s \le 1$). By applying the income identity to a closed economic system (Sushko et al. 2006), the following relationship can be established:

$$\begin{aligned} Y_t &= I_t + C_t \\ &= \nu(Y_{t-1} - Y_{t-2}) + \nu\mu(Y_{t-1} - Y_{t-2})^2 - \nu(Y_{t-1} - Y_{t-2})^3 + (1-s)Y_{t-1} + \varepsilon s Y_{t-2} + O^*(t) \end{aligned} \tag{9.3}$$

where $O^*(t) = I_{0t} + C_{0t}$. This denotes the sum of spontaneous investment and spontaneous consumption. $O^*(t)$ is assumed to depend only on time and to be independent of $Y_t$. According to Eq. (9.3), the corresponding continuous differential equation can be established. From a mathematical point of view, the evolution of a dynamical system can be described equally by a difference equation or a differential equation. This chapter employs the differential equation to study the periodic solution. In terms of the explicit expression of the periodic solution obtained, numerical analysis can easily be carried out.

Based on Eq. (9.3), the economic cycle evolution can be further formulated in an alternative way:

$$\begin{aligned} Y_{t+2} - 2Y_{t+1} + Y_t &= (\nu - 1 - s)(Y_{t+1} - Y_t) + \nu\mu(Y_{t+1} - Y_t)^2 \\ &\quad - \nu(Y_{t+1} - Y_t)^3 + (\varepsilon - 1)sY_t + O^*(t) \end{aligned} \tag{9.4}$$

By translating the discrete form of Eq. (9.4) into the continuous form (Li et al. 2007), the associated differential equation is established,

$$\ddot{Y}+(1+s-v)\dot{Y}-v\mu\dot{Y}^2+v\dot{Y}^3+(1-\varepsilon)sY=O^*(t) \tag{9.5}$$

Eq. (9.5) can express the response evolution of the economic cycle in a continuous way equivalently, as Eq. (9.3) does. Relying on the continuous differential equation, it is easier to employ some computational methods to analyze the response evolution of the economic cycle presented in this chapter.

In addition, this chapter simply considers the spontaneous function $O^*(t)$ as a cosine function to simulate a periodic inference on the economic evolution. $O^*(t)$ is mathematically expressed below:

$$O^*(t)=q\cos(\omega t) \tag{9.6}$$

where $q$ is the amplitude of the spontaneous adjustment and $\omega$ is the frequency of the spontaneous adjustment.

## 9.3 IMPLICIT HARMONIC BALANCE PROCEDURE

The implicit harmonic balance procedure was proposed by Cheung and Iu (1988) for analyzing periodic behaviors of some general dynamical systems, with quadratic and cubic nonlinearities in generalized displacement and viscous damping. Furthermore, the implicit harmonic balance procedure can also be applied to the relevant problems of multi-degrees-of-freedom nonlinear systems. This chapter extends the procedure of Cheung and Iu (1988) to study the economic cycle model with nonlinearity in damping. The nonlinearity in damping consists of linear, quadratic, and cubic terms. Due to these nonlinear terms, in addition to the primary response corresponding to the frequency of spontaneous adjustment, different harmonic responses occur, e.g., constant response, secondary response, and tertiary response. These additional harmonic responses are discussed in detail in this chapter.

At the steady state of the nonlinear economic cycle model, Eq. (9.5) has its periodic solution, which is assumed to be expressed by Fourier series, below:

$$Y(t)=\sum_{m=1}^{N}a^m cs(p_m\omega t) \tag{9.7}$$

where, for the index $m$, $a^m$ denotes the $m$th-term amplitude and $cs$ denotes the function of cosine or sine as follows

$$cs(p_m\omega t)=\begin{cases}\cos(p_m\omega t) & 1\le m\le N_c\\ \sin(p_m\omega t) & N_c<m\le N_c\end{cases} \tag{9.8}$$

where $p_m$ denotes a non-negative integer related to the $m$th term, $\omega$ is the frequency of the spontaneous function, $N=N_c+N_s$ is the sum of all the terms, $N_c$ denotes the number of cosine terms, and $N_s$ denotes the number of sine terms.

Similarly, the spontaneous function also can be expanded as follows:

$$O^*(t) = \sum_{m=1}^{N} q^m cs(p_m \omega t) \tag{9.9}$$

where $q^m$ is a known and constant amplitude of the spontaneous function.

In the proposed implicit harmonic balance procedure, each $p_m$ differs from the others for each cosine and sine subgroup. Furthermore, the number of higher harmonic terms should be enough in the periodic solution, in which some expected nonlinear responses can be obtained. If the periodic solution consists of both cosine and sine series, harmonic numbers should be the same for each subgroup, even in the cases where the harmonic number is null. The null in the sine series represents a trivial solution. A trial computation with more harmonic terms can be conducted to examine the improvement of the solution on the adequacy and the accuracy of the assumed solution for reflection of the nonlinear responses.

Substituting both Eq. (9.7) and Eq. (9.9) into Eq. (9.5) produces many products which can be expanded in the linear combinations of harmonic terms, according to trigonometric identities given in the Appendix. In terms of the same $p_m$, the coefficients of the corresponding harmonic terms are assembled and they equal to zero. This collection results in $N$ nonlinear algebraic equations. The variables include amplitudes $a^m$, frequency $\omega$, and the amplitude parameter $q^m$

$$f_i(A_1, \ldots, A_m, \ldots, A_N, \omega, q^m) = 0 \quad i = 1, \ldots, N \tag{9.10}$$

where $A_m$ corresponds to $a^m$ in the following manner:

$$A_m = a^m \quad m = 1, \ldots, N \tag{9.11}$$

Eq. (9.10) has a complicated explicit expression, which is generally very tedious and laborious to carry out, especially when a number of harmonic terms are used for the periodic solution. A straightforward and systematic procedure can be adopted to overcome this difficulty, through its implementation by computer.

Due to the complicated nonlinear functions and the presence of more harmonic terms, the explicit representations of Eq. (9.10) are difficult to derive. Therefore, some nonlinear equation solvers can fulfill this solution task using the residuals of these nonlinear equations. Evaluation of the function values $f_i$ in Eq. (9.10) can be conducted in an implicit way. It is necessary that the information is known for system Eq. (9.5) and the solution from Eq. (9.7).

The residuals of Eq. (9.10) can be obtained by expanding Eq. (9.5), term by term. Firstly, Eq. (9.7) is substituted into Eq. (9.5). Secondly, the products of the terms are expanded, using cosine and sine terms. Finally, the coefficients of these terms are collected with the same $p_m$. $N$ nonlinear algebraic equations can be formulated. Each term in the $i$th equation in Eq. (9.10) may contribute to the functions of the nonlinear economic cycle model, which is assembled according to Eqs. (9.7), (9.8), and (9.11).

For the $i$th nonlinear algebraic equation, the first and fifth terms in Eq. (9.5) can be formulated as Eq. (9.12) and Eq. (9.13), respectively

$$\ddot{Y} = \sum_{m=1}^{N} -p_m^2\omega^2 A_m cs(p_m\omega t) \tag{9.12}$$

$$(1-\varepsilon)sY = \sum_{m=1}^{N} (1-\varepsilon)sA_m cs(p_m\omega t) \tag{9.13}$$

Eq. (9.9) is expressed in a similar manner and is presented here again

$$O^*(t) = \sum_{m=1}^{N} q^m cs(p_m\omega t) \tag{9.14}$$

For the same cosine and sine functions, the coefficients of $cs(p_m\omega t)$ in Eqs. (9.12)–(9.14) are the components of the function $f_i$. For the damping terms, they contain the first derivative of $Y$, and the formulation procedure needs an indexing parameter for Eq. (9.10)

The linear damping of $\dot{Y}$ in Eq. (9.5) can be taken as

$$(1+s-v)\dot{Y} = \sum_{m=1}^{N} (1+s-v)A_m \frac{d}{dt} cs(p_m\omega t) \tag{9.15}$$

By the definition of $cs(p_m\omega t)$ in Eq. (9.8),

$$\frac{d}{dt} cs(p_m\omega t) = \begin{cases} -p_m\omega \sin(p_m\omega t) & 1 \le m \le N_c \\ p_m\omega \cos(p_m\omega t) & N_c < m \le N \end{cases} \tag{9.16}$$

Considering Eq. (9.16), $\alpha_m$ is adopted as an indexing parameter

$$\frac{d}{dt} cs(p_m\omega t) = \text{sign}(\alpha_m) p_m\omega cs(p_{|\alpha_m|}\omega t) \tag{9.17}$$

Therefore,

$$(1+s-v)\dot{Y} = \sum_{m=1}^{N} (1+s-v)A_m \text{sign}(\alpha_m) p_m\omega cs(p_{|\alpha_m|}\omega t) \tag{9.18}$$

According to $\alpha_m$, the sign of these components can be determined as well as the affiliated harmonic terms. After that, the corresponding coefficient of $cs(p_{|\alpha_m|}\omega t)$ takes part in the function $f_{|\alpha_m|}$.

For the quadratic term of $\dot{Y}$ in Eq. (9.5), $\alpha_m$ and $\beta_n$ are needed as two indexing parameters. They are used to allocate the derivatives of the quadratic term presented by Eq. (9.19). Referring to the Appendix, the quadratic term in Eq. (9.19) is further expanded into the linear combination of two components. Herein, $\mu_{rmn}$ is needed as one additional indexing parameter. It presents the information of the two components. The coefficients of $cs(p_{|\alpha_m|}\omega t)$ are allocated into the associated $f_{|\mu_{rmn}|}$ and the quadratic term is treated by harmonic balance.

$$
\begin{aligned}
-v\mu\dot{Y}^2 &= \sum_{m=1}^{N}\sum_{n=1}^{N} -v\mu A_m A_n \times \frac{d}{dt}cs(p_m\omega t)\frac{d}{dt}cs(p_n\omega t) \\
&= \sum_{m=1}^{N}\sum_{n=1}^{N} -v\mu A_m A_n p_m p_n \omega^2 \\
&\quad \times sign(\alpha_m)sign(\beta_n)cs(p_{|\alpha_m|}\omega t)cs(p_{|\beta_n|}\omega t) \qquad (9.19) \\
&= \sum_{m=1}^{N}\sum_{n=1}^{N}\sum_{s=1}^{2} \frac{1}{2}(-v\mu)_m A_n p_m p_n \omega^2 \\
&\quad \times sign(\mu_{smn})sign(\alpha_m)sign(\beta_n)cs(p_{|\mu_{smn}|}\omega t)
\end{aligned}
$$

For the cubic term of $\dot{Y}$ in Eq. (9.5), the manipulation procedure is similar. Four indexing parameters are used for three derivative terms. The linear combinations of harmonic components are established. The trigonometric identities are also given in the Appendix. The coefficients of $cs(p_{|\mu_{smn}|}\omega t)$ are assigned into the corresponding $f_{|\eta_{smnr}|}$, according to $\eta_{smnr}$.

$$
\begin{aligned}
v\dot{Y}^3 &= \sum_{m=1}^{N}\sum_{n=1}^{N}\sum_{r=1}^{N} vA_m A_n A_r \times \frac{d}{dt}cs(p_m\omega t)\frac{d}{dt}cs(p_n\omega t)\frac{d}{dt}cs(p_r\omega t) \\
&= \sum_{m=1}^{N}\sum_{n=1}^{N}\sum_{r=1}^{N} vA_m A_n A_r \times p_m p_n p_r \omega^3 sign(\alpha_m)sign(\beta_n)sign(\gamma_r) \\
&\quad \times cs(p_{|\alpha_m|}\omega t)cs(p_{|\beta_n|}\omega t)cs(p_{|\gamma_r|}\omega t) \qquad (9.20) \\
&= \sum_{m=1}^{N}\sum_{n=1}^{N}\sum_{r=1}^{N}\sum_{s=1}^{4} \frac{1}{4} vA_m A_n A_r \times p_m p_n p_r \omega^3 sign(\eta_{smnr})sign(\alpha_m) \\
&\quad \times sign(\beta_n)sign(\gamma_r)cs(p_{|\eta_{smnr}|}\omega t)
\end{aligned}
$$

Finally, the nonlinear algebraic equations, $f_i(A_1,\ldots,A_m,\ldots,A_N,\omega,q^m), i=1,\ldots,N$, are established. There are $N+1$ unknown variables, except $q^m$ is usually a known constant. Therefore, one of the variables is first kept constant and the others are solved. Once the solution is achieved, the constant variable can be increased a little. When the solution has a smooth enough path, the next solution can be extrapolated from the solution obtained by a simple Lagrangian formula. The arc length $\delta$ is selected to control the extrapolation procedure. The vector **A**, containing $A_1,\ldots,A_m,\ldots,A_N$ and $\omega$ approximately, is given below:

$$\{A(\delta)\} \cong \sum_{k=0}^{r}\left[\prod_{i=0,i\neq k}^{r}\left(\frac{\delta-\delta_i}{\delta_k-\delta_i}\{A(\delta_k)\}\right)\right] \tag{9.21}$$

When $r=2$, a quadratic extrapolation procedure is used

$$\delta_0=0,\ \ \delta_k=\delta_{k-1}+|\{A(\delta_k)\}-\{A(\delta_{k-1})\}| \tag{9.22}$$

The initial trial solution is recommended with the small-amplitude solutions of a linear system, with free or forced vibration. For tracing the response curve along with the frequency, the initial frequency solution is chosen in the region with the occurrence of small amplitudes.

## 9.4 NUMERICAL ANALYSIS

Numerical analysis is conducted to examine the periodic response of the model Eq. (9.5). The parameters in Eqs. (9.5) and (9.6) are given: $\nu=0.2$, $s=0.2$, $\mu=1$, $\varepsilon=0.05$, and $q=0.8$.

The periodic solution of Eq. (9.5) is assumed to be as follows:

$$\begin{aligned} Y &= A_1\cos(p_1\omega t)+A_2\cos(p_2\omega t)+A_3\cos(p_3\omega t)+A_4\cos(p_4\omega t)\\ &= A_5\cos(p_5\omega t)+A_6\cos(p_6\omega t)+A_7\cos(p_7\omega t)+A_8\cos(p_8\omega t)\end{aligned} \tag{9.23}$$

The selection criteria on $p_m$ follow that, if the periodic solution consists of both cosine and sine series, the assigned harmonic numbers in each subgroup are the same. The non-negative integers are defined as follows: $p_1=p_5=0$, $p_2=p_6=1$, $p_3=p_7=2$, $p_4=p_8=3$. By letting $p_1=p_5=0$, the constant of the periodic solution is examined. Although $p_5=0$ represents a trivial solution in the sine series, $p_5$ is still adopted to satisfy the selection criteria. By letting $p_2=p_6=1$, the primary response of the model Eq. (9.5) is investigated. A case with $p_3=p_7=2$ is devoted to examining the effects of the quadratic term. Similarly, by setting $p_4=p_8=3$, the effects of the cubic term are studied. Eq. (9.23) can be expressed in an alternative manner as shown below:

$$\begin{aligned} Y &= A_1+\sqrt{A_2^2+A_6^2}\cos(\omega t+\phi_1)+\sqrt{A_3^2+A_7^2}\cos(2\omega t+\phi_2)+\sqrt{A_4^2+A_8^2}\cos(3at+\phi_3)\\ &=\bar{A}_0+\bar{A}_1\cos(\omega t+\phi_1)+\bar{A}_2\cos(2\omega t+\phi_2)+\bar{A}_3\cos(3at+\phi_3)\end{aligned} \tag{9.24}$$

where $\bar{A}_0=A_1$; $\bar{A}_1=\sqrt{A_2^2+A_6^2}$; $\bar{A}_2=\sqrt{A_3^2+A_7^2}$; $\bar{A}_3=\sqrt{A_4^2+A_8^2}$; $\phi_1$, $\phi_2$, $\phi_3$ are phase angles. It is clear to investigate the effects of different nonlinearity terms, using $\bar{A}_0$, $\bar{A}_1$, $\bar{A}_2$, $\bar{A}_3$. Furthermore, the numbers in the subscript of each variable denote only the non-negative integers: $p_m = 0, 1, 2, 3$, respectively. They are studied in detail in the following subsection of this chapter.

### 9.4.1 The Comparison of the Periodic Solution with the Simulation Result

First, the effectiveness of the extended implicit harmonic balance (IHB) procedure is investigated. The periodic solution is compared with the simulation results (i.e., Monte Carlo simulation, MCS). In Figure 9.1a, IHB denotes the total amplitude along with the increase of $\omega$ in terms of Eq. (9.23), which is solved by the extended implicit harmonic balance procedure. MCS denotes the total amplitude given by Monte Carlo simulation, according to Eq. (9.5). The response samples are obtained by the fourth-order Runge-Kutta algorithm. The total amplitude for each $\omega$ is taken as the steady state, which is far from the initial state. Figure 9.1a shows close agreement between IHB and MCS, along with the increase of $\omega$. Furthermore, two typical vibratory processes are examined. One vibratory process is taken when $\omega = 0.0098$. This frequency is located around the resonance region. The other vibratory process is taken when $\omega = 4.0003$. Because these two frequencies differ markedly in their magnitude, the time durations are taken differently to obtain the steady periodic response. Figures 9.1b and c show that, in the beginning, the non-steady responses occur and difference exists between IHB and MCS. As time passes, the response reaches the steady state. Close agreement between IHB and MCS is achieved, which verifies the effectiveness of the extended implicit harmonic balance procedure. It is also interesting that, when the frequency of harmonic spontaneous adjustment is located in the resonance region, the nonlinear economic cycle model vibrates in an almost symmetric manner, as shown in Figure 9.1b. However, if the frequency is far from the resonance region, the vibratory response has a non-zero mean, which implies that a constant exists in the periodic response of the economic cycle. This phenomenon is shown in Figure 9.1c.

### 9.4.2 The Periodic Solution of the Nonlinear Economic Cycle Model

Figure 9.2 provides the periodic solution with its amplitudes, along with the frequency $\omega$, which is obtained by the extended implicit harmonic balance procedure. Figure 9.2a shows the evolution of the constant response $\bar{A}_0$.$\bar{A}_0$ first gradually increases as the frequency increases from zero. The maximum peak of $\bar{A}_0$ is achieved when $\omega$ reaches 0.4573. This value is around the natural frequency of the corresponding linear economic cycle model (i.e., Eq. (9.5)) without all the damping terms and $\sqrt{(1-\varepsilon)s} = 0.4359$. After the peak is achieved, $\bar{A}_0$ rapidly decreases as $\omega$ increases further, and becomes very small when $\omega$ is very large. In a word, the resonance mechanism exists in the constant response.

By contrast, the primary response $\bar{A}_1$ behaves in a different way, as shown in Figure 9.2b. The maximum magnitude of $\bar{A}_1$ corresponds to $\omega = 0$. Subsequently, the

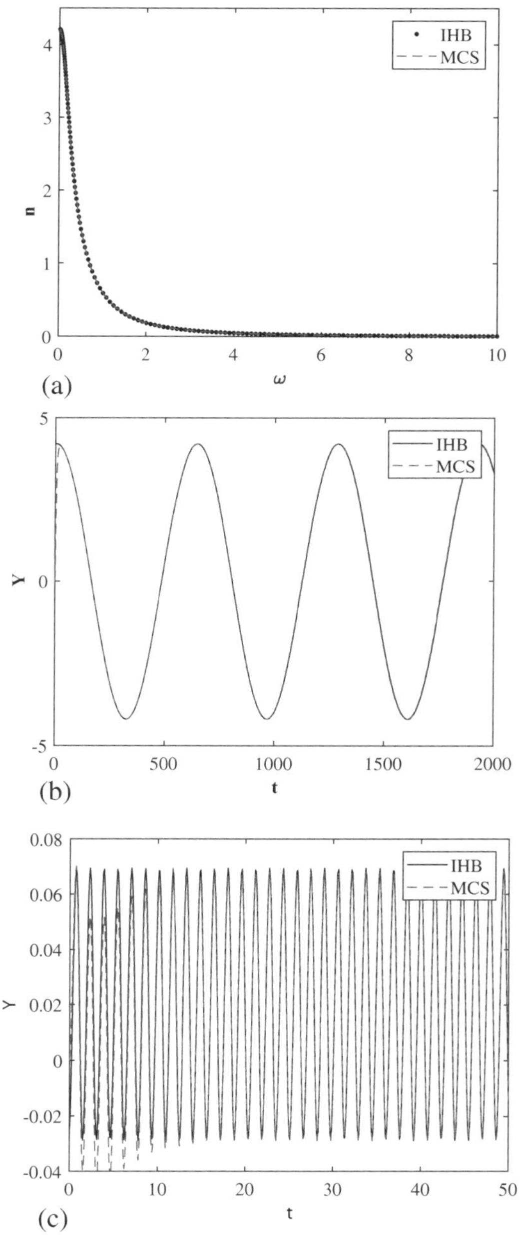


**FIGURE 9.1** Comparison of the solutions of the implicit harmonic balance procedure (IHB) with the Monte Carlo simulation (MCS) results: (a) Amplitude, (b) vibratory response at $\omega = 0.0098$, and (c) vibratory response at $\omega = 4.0003$.

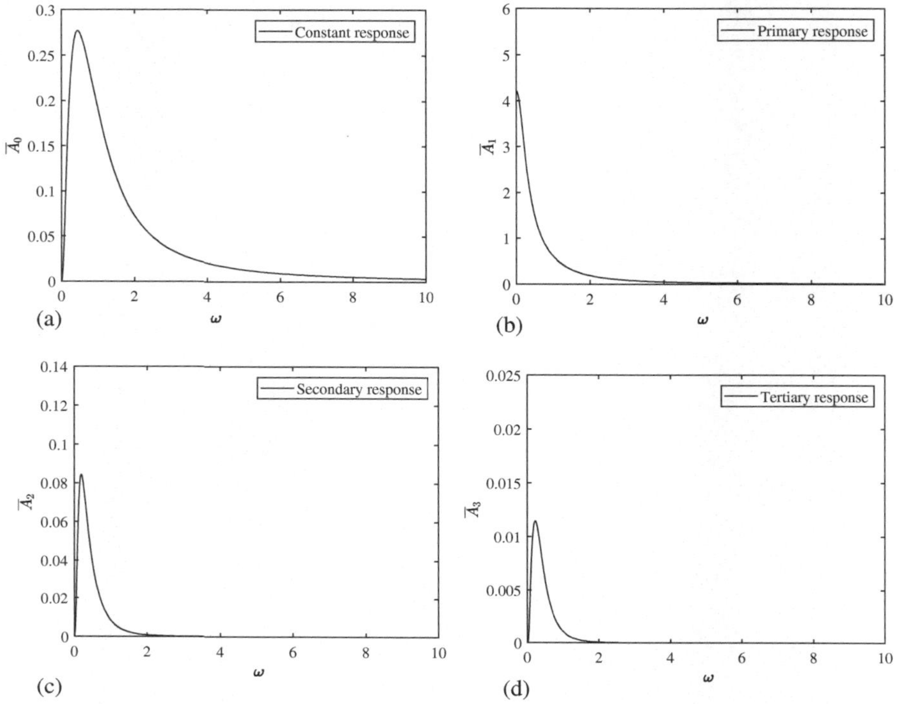


**FIGURE 9.2** The amplitudes of the periodic solution: (a) Constant response, (b) primary response, (c) secondary response, and (d) tertiary response.

magnitude of $\bar{A}_1$ rapidly decreases in response to increasing $\omega$. For the secondary response, $\bar{A}_2$, and the tertiary response, $\bar{A}_3$, in Figures 9.2c and 9.2d, respectively, their evolution is just like the one for the constant response, $\bar{A}_0$. In response to the increase of $\omega$ from zero, their magnitudes gradually increase, reach the peaks, and then decrease rapidly. The peak of $\bar{A}_2$ corresponds to $\omega = 0.1908$ and the peak of $\bar{A}_3$ corresponds to $\omega = 0.2234$.

Comparison of the magnitudes among $\bar{A}_0, \bar{A}_1, \bar{A}_2$, and $\bar{A}_3$ shows that the primary response $\bar{A}_1$ is much higher than the other responses. As higher harmonic responses are considered, the effects become much smaller.

### 9.4.3 The Effects of the Quadratic Term on the Periodic Solution

Finally, the effects of the quadratic term in Eq. (9.5) are examined by using different magnitudes of $\mu$. Figure 9.3 presents a comparison of the magnitudes of different responses when $\mu$ increases from 1 to 3. Figure 9.3a presents that $\bar{A}_0$ increases as $\mu$ increases. However, the primary response, $\bar{A}_1$, exhibits little change as $\mu$ changes, which implies that $\bar{A}_1$ is not affected by the quadratic term. For $\bar{A}_2$, its behavior is

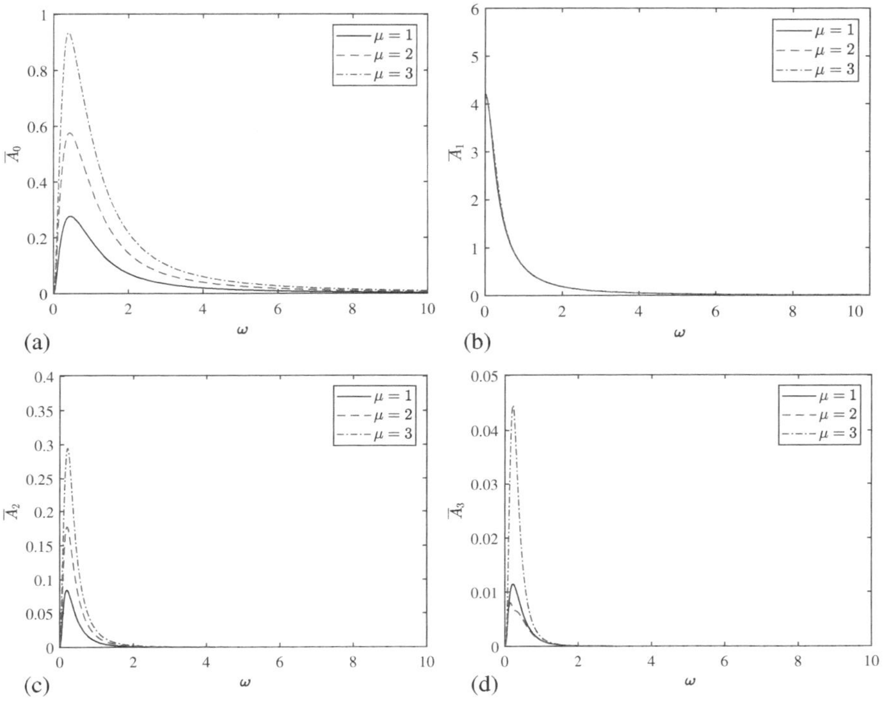


**FIGURE 9.3** The amplitudes of the periodic solution for different values of $\mu$: (a) Constant response, (b) primary response, (c) secondary response, and (d) tertiary response.

just like that of $\overline{A}_0$. The magnitude of $\overline{A}_2$ increases along with the increase in $\mu$. After the peak is achieved, $\overline{A}_2$ decreases steadily in response to increasing $\omega$. However, this is not the case for $\overline{A}_3$. The magnitude of $\overline{A}_3$ first decreases when $\mu = 2$ and then decreases when $\mu = 1$. When $\mu = 3$, the magnitude of $\overline{A}_3$ increases more than when $\mu = 1$. These observations show that the quadratic term has complicated effects on different harmonic responses.

## 9.5 CONCLUSIONS

This chapter described an implicit solution procedure of harmonic balance for analyzing the response of nonlinear economic cycle models with harmonic spontaneous adjustment. Firstly, the economic cycle model is derived, consisting of a linear-quadratic-cubic function of the difference between pre-income and prior income. The function is used to model a general effect of incomes on investment. Secondly, the implicit harmonic balance procedure is applied to the developed model. A subsequent numerical analysis is conducted. The periodic solution of the nonlinear economic cycle model is investigated and verified in terms of its constant response, primary response, secondary response, and tertiary response. These responses behave in significantly different manners. Generally, the primary response is much higher than the other responses. As higher harmonic responses are considered, the effects become much smaller. A further parametric study shows that the presence of the nonlinear terms leads to the occurrence of different harmonic dynamic responses. When the frequency of harmonic spontaneous adjustment is located in the resonance region, the nonlinear economic cycle model vibrates in an almost symmetrical manner. However, if the frequency is far from the resonance region, the vibratory response has a non-zero mean, which implies that a constant exists in the periodic response of the economic cycle. In addition, along with the increase in the quadratic term, the constant response and the secondary response also increase correspondingly. However, the primary response shows little change and the tertiary response decreases in the case of $\mu=2$ and increases in the case of $\mu=3$. These observations show that the nonlinear terms have complex effects on different harmonic responses.

## APPENDIX

### Trigonometric Identities

$$\begin{aligned} c_i c_j &= \frac{1}{2}(c_{i+j} + c_{i-j}) \\ c_i s_j &= \frac{1}{2}(s_{i+j} - s_{i-j}) \\ s_i s_j &= \frac{1}{2}(-c_{i+j} + c_{i-j}) \end{aligned} \tag{9.25}$$

$$\begin{aligned} c_i c_j c_k &= \frac{1}{4}(c_{i+j+k} + c_{i+j-k} + c_{i-j+k} + c_{i-j-k}) \\ c_i c_j s_k &= \frac{1}{4}(s_{i+j+k} - s_{i+j-k} + s_{i-j+k} - s_{i-j-k}) \\ c_i s_j s_k &= \frac{1}{4}(-c_{i+j+k} + c_{i+j-k} + c_{i-j+k} - c_{i-j-k}) \\ s_i s_j s_k &= \frac{1}{4}(-s_{i+j+k} + s_{i+j-k} + s_{i-j+k} - s_{i-j-k}) \end{aligned} \tag{9.26}$$

where $c_i = \cos(p_i \omega t)$ and $s_i = \sin(p_i \omega t)$.

## REFERENCES

Amman, H. 1996. Numerical methods for linear-quadratic models. In *Handbook of Computational Economics* Vol. 1, North-Holland, Amsterdam, Chap. 13:587–618.

Bashkirtseva, I., L. Ryashko, and A. Sysolyatina. 2016. Analysis of stochastic effects in Kaldor-type business cycle discrete model. *Commun Nonlinear Sci Numer Simul* 36:446–456.
Bashkirtseva, I., D. Radi, L. Ryashko, and T. Ryazanova. 2018a. On the stochastic sensitivity and noise-induced transitions of a Kaldor-type business cycle model. *Comput Econ* 51:699–718.
Bashkirtseva, I., L. Ryashko, and T. Ryazanova. 2018b. Stochastic sensitivity analysis of the variability of dynamics and transition to chaos in the business cycles model. *Commun Nonlinear Sci Numer Simul* 54:174–184.
Cheung, Y.K., and V.P. Iu. 1988. An implicit implementation of harmonic balance method for non-linear dynamic systems. *Eng Comput* 5:134–140.
Chian, A.C.L., F.A. Borotto, E.L. Rempel, and C. Rogers. 2005. Attractor merging crisis in chaotic business cycles. *Chaos Solitons Fractals* 24:869–875.
Dai, H., X. Yue, J. Yuan, and D. Xie. 2018. Dealiasing harmonic balance method for obtaining periodic solutions of an aeroelastic system. *Aerosp Sci Technol* 77:244–255.
Detroux, T., L. Renson, L. Masset, and G. Kerschen. 2015. The harmonic balance method for bifurcation analysis of large-scale nonlinear mechanical systems. *Comput Method Appl Mech Eng* 296:18–38.
Foupouapouognigni, O., C. Nono Dueyou Buckjohn, M. Siewe Siewe, and C. Tchawoua. 2017. Nonlinear electromechanical energy harvesters with fractional inductance. *Chaos Solitons Fractals* 103:12–22.
Goodwin, R.M. 1951. The nonlinear accelerator and the persistence of business cycles. *Econometrica* 19:1–17.
Grolet, A., and F. Thouverez. 2015. Computing multiple periodic solutions of nonlinear vibration problems using the harmonic balance method and Groebner bases. *Mech Syst Signal Process* 52–53:529–547.
Guerrazzi, M., and M. Sodini. 2018. Efficiency-wage competition and nonlinear dynamics. *Commun Nonlinear Sci Numer Simul* 58:62–77.
Hattaf, K., D. Riad, and N. Yousfi. 2017. A generalized business cycle model with delays in gross product and capital stock. *Chaos Solitons Fractals* 98:31–37.
Li, J., and S. Li. 2015. Dynamics of a nonlinear business cycle model under Poisson white noise excitation. *J Syst Sci Inf* 3:176–183.
Li, W., W. Xu, J. Zhao, and Y. Jin. 2007. Stochastic stability and bifurcation in a macroeconomic model. *Chaos Solitons Fractals* 31:702–711.
Lin, Z., J. Li, and S. Li. 2016. On a business cycle model with fractional derivative under narrow-band random excitation. *Chaos Solitons Fractals* 87:61–70.
Liu, X., W. Cai, J. Lu, and Y. Wang. 2015. Stability and Hopf bifurcation for a business cycle model with expectation and delay. *Commun Nonlinear Sci Numer Simul* 25:149–161.
Matsumoto, A., U. Merlone, and F. Szidarovszky. 2018. Goodwin accelerator model revisited with fixed time delays. *Commun Nonlinear Sci Numer Simul* 58:233–248.
Puu, T., and I. Sushko. 2004. A business cycle model with cubic nonlinearity. *Chaos Solitons Fractals* 19:597–612.
Riad, D., K. Hattaf, and N. Yousfi. 2016. Dynamics of a delayed business cycle model with general investment function. *Chaos Solitons Fractals* 85:110–119.
Serrano, S., R. Barrio, A. Dena, and M. Rodrıguez. 2012. Crisis curves in nonlinear business cycles. *Commun Nonlinear Sci Numer Simul* 17: 788–794.
Sushko, I., T. Puu, and L. Gardini. 2006. A Goodwin-type model with cubic investment function. In *Business Cycle Dynamics: Models and Tools*, Springer, Berlin Heidelberg, Chap. 11:299–316.
Tran, K., G. Yin, and L.Y. Wang. 2016. A generalized Goodwin business cycle model in random environment. *J Math Anal Appl* 438:311–327.

Xu, Y., A.C.J. Luo, and Z. Chen. 2017. Analytical solutions of periodic motions in 1-dimensional nonlinear systems. *Chaos Solitons Fractals* 97:1–10.

Yao, W., and S. Marques. 2018. A harmonic balance method for nonlinear fluid structure interaction problems. *Comput Struct* 201:26–36.

Zhao, M., and C. Li. 2016. Complex dynamic behavior of an economic cycle model. *J Differ Equation Appl* 22:1777–1790.

# 10 Response Evolution of a Marine Riser in Random Sea Waves

*Haitao Zhu, Guoqian Geng, Yang Yu, and Lixin Xu*

## CONTENTS



## 10.1 INTRODUCTION

Since the beginning of the 21st century, technology and economy have entered a stage of rapid development. In order to meet the demand for resources, people have set their sights on the ocean. There are abundant oil and gas resources, biological resources, and renewable resources in the ocean. However, according to the current data, due to the complexity of the seabed environment and the limitation of technology and equipment, the exploration rate and overall exploitation degree of marine oil and gas resources are relatively low. In the process of marine oil and gas resource exploitation, an important item of equipment is a marine riser. The marine riser is the connecting pipeline connecting between the platform and the subsea wellhead and is also the transportation channel for oil and gas. It has been applied to different types of platforms. At present, most studies are about top tension risers, steel catenary risers, and flexible risers.

This chapter focuses on the behavior of the top tension riser. When the riser is subjected to sea waves, the motion of heave and surge occurs. The riser will be tensioned or compressed, resulting in the axial changing force along the marine riser. Meanwhile, the riser also vibrates horizontally along with the sea waves. The two types of motion lead to parametric vibration. These loads may cause the riser to collide, resulting in vortex-induced vibration, parametric vibration, etc., thereby causing fatigue damage or fracture damage to the riser (He and Low 2012; Thorsen et al. 2019). Parametric vibration of the marine riser has strong nonlinearity, which requires the combination of a variety of dynamic knowledge to solve the nonlinear problem. Many researchers have carried out relevant research on parametric vibration of marine structures. In different fields, researchers have used different methods and focused on studying nonlinear vibration. Comparing the results of model tests and numerical simulations can better study the dynamic characteristics of marine risers (Yin et al. 2019). Comparing the Mathieu instability diagram, obtained by the Mathieu equation, with the numerical simulation results, based on Floquet theory, can obtain the analysis results on the stability of the marine riser (Zhang and Tang 2014). Patel and Park (1995) combined the Romberg method with the Runge-Kutta method for dealing with the parametric vibration of the marine riser. In recent years, the probability density function (PDF) method has been commonly used for studying nonlinear vibration (Zhu 2017; Han et al. 2016). In addition, Cabrera-Miranda and Paik (2017) conducted a study on the probabilistic distribution of excitations acting on marine risers. This helps to determine the parameters in a nominal design. They developed a metamodel to determine the loads in a form of multiple input parameters. The results showed that the loads on marine risers have highly random characteristics. Ni et al. (2018) performed the stochastic analysis of marine risers with material uncertainties in mass density and elastic modulus. A stochastic finite element method was used along with the model reduction technique. Ni et al. (2019) carried out a stochastic study on the vibration of marine risers. They considered not only fluid–structure interaction but also system uncertainties. A particle finite element method was used to simulate the fluid behavior. A finite element method was adopted to study the behavior of offshore structures. Alfosail and Younis (2019) investigated an inclined marine riser with 3:1 internal resonance. They considered the riser flexural rigidity, axial force, and nonlinear geometry, as well as initial static deflection. Yang et al. (2020) studied deep-water marine riser systems. Flexible multibody dynamics were used in their study. The stiffness of the auxiliary lines has a remarkable effect on the tension safety of marine risers.

On the other hand, many practical problems can be modeled by differential equations in mathematics, mechanics, physics, economics, etc. Therefore, these differential equations are increasingly applied to the relevant fields. The solutions of differential equations can be exactly or approximately solved by some methods. Among these methods, a path integration method is widely used for its simplicity and efficiency. The idea of path integration is to discretize the solution in space and in time, and to replace the integral with the sum of paths. The result is that the short-term transition PDF can be obtained by connecting and gradually accumulating the integration time. In the beginning, path integration was applied to the field of quantum mechanics, and then it was extended to a variety of research fields. At the beginning, it was mainly used in the research fields of physics (Grosjean and Goovaerts 1973) and quantum mechanics (Barvinsky 1998). With the development of science and technology, it has been applied to more and more important fields. Based on Gauss-Legendre integration scheme, Yu et al. (1997) proposed a path integration method to calculate the response PDF of the system driven by Gaussian white noise. The Gauss-Legendre scheme was employed for the integral. This method has been widely used in the research of later scholars. Xie et al. (2006) studied the PDF of a Duffing-Rayleigh oscillator by path integration. The two types of excitation (i.e., harmonic and random) were considered simultaneously. The Gauss-Legendre integration scheme was used in their study. Narayanan and Kumar (2012) improved the path integration method to study the stochastic and chaotic responses of some nonlinear systems. In recent years, some new advances in path integration methods have been made. Alevras and Yurchenko (2016) proposed a graphics processing unit (GPU) computing approach to perform path integration. A stationary joint PDF was investigated for a dynamic oscillator under random excitation. Zhu and Duan (2016) adopted path integration to investigate the PDF of nonlinear random ship roll motion. In their study, the interpolation is in terms of the Gauss–Legendre integration scheme. The short-time transition PDF is approximated by Gaussian PDF. Gaidai et al. (2017) studied the extreme statistics of a random Jeffcott rotor uniaxially driven by white noise. Fast Fourier transform (FFT) was applied to implement the path integration algorithm. The computational time was significantly reduced. Chai et al. (2017) conducted a comparative study between path integration and stochastic averaging. The illustrative examples were about nonlinear random roll motion. Gaidai et al. (2019) integrated a GPU technique into path integration to analyze the random response of a rotating shaft driven by colored noise. Hasnijeh et al. (2019) employed an adaptive time-stepping path integration to examine the random response of a nonlinear time-varying spur gear model. Ren and Xu (2020) improved a path integration method in the case of nonlinear oscillators subjected to Poisson impulses.

This chapter is concerned with the joint PDF evolution of a parametrically excited marine riser. Firstly, the load of the marine riser is analyzed and the vibration model is established. Secondly, the problem of the marine riser is simplified, and the equation of motion is established for the marine riser. Thirdly, the Galerkin method is used to obtain the first-order mode vibratory differential equation. After that, a path integration method is used to study the response. Its effectiveness is verified by comparing the PDFs among path integration, equivalent linearization, and numerical simulation. A few time events are selected from the initial state to the stationary state to show the joint PDF evolution process. Lastly, the stationary PDFs solved by path integration are compared among all the cases.

## 10.2 MARINE RISER SYSTEM

In this chapter, to simplify the problem, some assumptions are made. The sea waves go along the same direction. The geometry and material properties are constant along the riser length. The study considers not only hydrodynamic damping but also the nonlinearity due to the riser curvature. The simplified diagram is given for a marine riser in Figure 10.1.

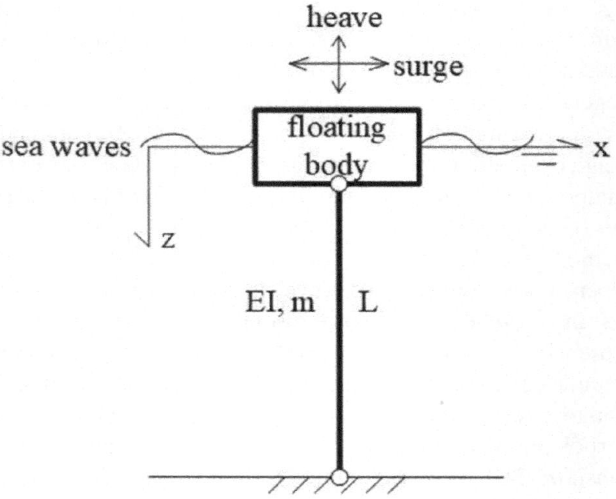


**FIGURE 10.1** Schematic representation of the motion of a marine riser.

For a marine riser, Zhang and Tang (2014) proposed the equation of motion below:

$$\begin{aligned} &\mathrm{EI}\frac{\partial^4 w(z,t)}{\partial z^4} - \frac{3}{2}EI\frac{\partial^2}{\partial z^2}\left[(\frac{\partial w(z,t)}{\partial z^2})^2\left|\frac{\partial^2 w(z,t)}{\partial z^2}\right|\right] \\ &-[T - f_2(t)]\frac{\partial^2 w(z,t)}{\partial z^2} + \alpha\frac{\partial w(z,t)}{\partial t} + \overline{m}\frac{\partial^2 w(z,t)}{\partial^2 t} = f_1(z,t) \end{aligned} \tag{10.1}$$

The definitions of the above variables for the marine riser are as follows: $w(z,t)$ represents the horizontal displacement, $f_1(z,t)$ is the external excitation, $f_2(t)$ is the parametric excitation, and $z$ denotes the vertical coordinate. In addition, $EI$ denotes the bending stiffness, $T$ is the static tension, $\alpha$ denotes the damping coefficient, and $\overline{m}$ denotes the equivalent mass per unit length.

Adopting the Galerkin method and letting $w(z,t) = \phi(z)u(t)$, Eq. (10.2) represents the differential equation describing the parametric vibration for the riser.

$$\ddot{u} + c\dot{u} + k_1 u + k_2 u^3 = \mu W_1(t) + \varepsilon W_2(t)u \tag{10.2}$$

where $\phi(z)$ denotes the mode shape function; $u(t)$ denotes the coordinate function. This study considers only the first vibratory mode, i.e., $\phi(z) = \sin\left(\frac{\pi z}{L}\right)$. Herein $L$ denotes the length, $\ddot{u}$ denotes acceleration, $\dot{u}$ denotes velocity, and $u$ denotes displacement. Some new variables are introduced to obtain a dimensionless equation

$$c=\frac{\alpha}{m}\quad \mu=\frac{\pi}{2mL}\quad \varepsilon=\frac{\pi^2}{mL^2}\quad k_1=\frac{EI\pi^4}{mL^4}+\frac{\pi^2 T}{mL^2}$$
$$k_2=\frac{3EI\pi^6}{L^6}\quad W_1(t)=\int_o^L f_1(z,t)dz\quad W_2(t)=f_2(t) \tag{10.3}$$

In addition, $W_1(t)$ and $W_2(t)$ are Gaussian white noises and they are correlated as follows.

$$E[W_1(t)W_1(t+\tau)]=I_{11}\delta(\tau)$$
$$E[W_2(t)W_2(t+\tau)]=I_{22}\delta(\tau) \tag{10.4}$$
$$E[W_1(t)W_2(t+\tau)]=I_{12}\delta(\tau)\quad I_{12}=r\sqrt{I_{11}I_{22}}$$

where $r$ is the correlation coefficient, $E[*]$ is the expectation of [*], $I_{11}$ is the intensity of $W_1(t)$, $I_{22}$ is the intensity of $W_2(t)$, and $\delta(t)$ denotes the Dirac delta function.

## 10.3 PATH INTEGRATION PROCEDURE

Letting $x=u$ and $v=\dot{u}$, Eq. (10.2) can be transformed into Eq. (10.5).

$$\begin{cases}\dot{x}=v\\ \dot{v}=-cv-k_1x-k_2x^3+\mu W_1(t)+\varepsilon W_2(t)x\end{cases} \tag{10.5}$$

Then, the moment-closure equations can be obtained using Gaussian closure. In terms of the Gaussian assumption, the cumulants that are higher than the second order are null. In Eq. (10.6), $\dot{m}_{ij}$ is the derivative of $m_{ij}$ with respect to time.

$$\begin{cases}\dot{m}_{10}=m_{01}\\ \dot{m}_{01}=-cm_{01}-k_1m_{10}+2k_2m_{10}^3-3k_2m_{10}m_{20}\\ \dot{m}_{20}=2m_{11}\\ \dot{m}_{11}=m_{02}-cm_{11}-k_1m_{20}-3k_2m_{20}^2+2k_2m_{10}^4\\ \dot{m}_{02}=-2cm_{02}-2k_1m_{11}+4k_2m_{01}m_{10}^3-6k_2m_{11}m_{20}+\mu^2I_{11}+\varepsilon^2I_{22}m_{20}+2I_{12}\mu\varepsilon m_{10}\end{cases} \tag{10.6}$$

The Runge-Kutta method with an initial value is employed for the solution to Eq. (10.6). The solution is solved step by step in terms of the known solution at the previous step at $t=t_{i-1}$. The initial state ($x=x_{kl}^{(i-1)}$, $v=v_{m'n'}^{(i-1)}$) is given below:

$$m_{10}(t_{i-1})=x_k^{(i-1)},\quad m_{01}(t_{i-1})=v_l^{(i-1)},\quad m_{11}(t_{i-1})=x_k^{(i-1)}v_l^{(i-1)},$$
$$m_{20}(t_{i-1})=(x_k^{(i-1)})^2,\quad m_{02}(t_{i-1})=(v_l^{(i-1)})^2 \tag{10.7}$$

Yu et al. (1997) proposed the Gauss-Legendre scheme to implement path integration for obtaining PDFs in this chapter. Two Gaussian points are used in each integration interval. The two-dimensional expression of PDF is given below.

$$p(x_r^i, v_s^i, t_i) = \frac{\Delta x}{2}\frac{\Delta v}{2}\sum_{k=1}^{2n}\sum_{l=1}^{2m} p(x_k^{(i-1)}, v_l^{(i-1)}, t_{(i-1)}) \times q(x_r^i, v_s^i, t_i \,|\, x_k^{(i-1)}, v_l^{(i-1)}, t_{(i-1)}) \quad (10.8)$$

where $m$ is the subinterval number in the $x$ direction, $n$ is the subinterval number in the $v$ direction, $\Delta x$ is the subinterval length in the $x$ direction, $\Delta v$ is the subinterval length in the $v$ direction, and $(x_r, v_s)$ is the coordinate of the Gaussian points.

In each subinterval with $(x_r, x_l)$ and $(v_r, v_l)$, the coordinates of the two Gaussian points are given:

$$\begin{cases} x_1 = x_l + 0.211375(x_r - x_l) \\ x_2 = x_r - 0.211375(x_r - x_l) \end{cases} \qquad \begin{cases} v_1 = v_l + 0.211375(v_r - v_l) \\ v_2 = v_r - 0.211375(v_r - v_l) \end{cases} \quad (10.9)$$

As described above, the short-time transition PDF is a joint Gaussian PDF, below:

$$q(x_r^i, v_s^i, t_i \,|\, x_k^{(i-1)}, v_l^{(i-1)}, t_{(i-1)}) = \frac{1}{2\pi\sigma_1(t_i)\sigma_2(t_i)\sqrt{1-\rho^2(t_i)}} \times \exp\left\{\frac{-1}{2(1-\rho^2(t_i))} \times \left[\frac{(x_r^{(i)} - m_{10}(t_i))^2}{\sigma_1^2(t_i)} - \frac{2\rho(t_i)(x_r^{(i)} - m_{10}(t_i))}{\sigma_1(t_i)\sigma_2(t_i)} \times (v_s^{(i)} - m_{01}(t_i)) + \frac{(v_s^{(i)} - m_{01}(t_i))^2}{\sigma_2^2(t_i)}\right]\right\} \quad (10.10)$$

$$\begin{cases} \sigma_1^2(t_i) = m_{20}(t_i) - \left[m_{10}(t_i)\right]^2 \\ \sigma_2^2(t_i) = m_{02}(t_i) - \left[m_{01}(t_i)\right]^2 \\ \rho(t_i) = \left[m_{11}(t_i) - m_{10}(t_i)m_{01}(t_i)\right] / \left[\sigma_1(t_i)\sigma_2(t_i)\right] \end{cases}$$

where $m_{ij} = E\left[x^i v^j\right]$

## 10.4 NUMERICAL ANALYSIS

Three typical examples are chosen to study not only the stationary case but also the non-stationary case. The effectiveness of the adopted method is examined by comparing the PDFs obtained by different methods, i.e., equivalent linearization and numerical simulation. The details of each example are shown in Table 10.1.

**TABLE 10.1**
**The Values of the Parameters in Numerical Analysis in the Three Cases**

| | $c$ | $k_1$ | $k_2$ | $\mu$ | $\varepsilon$ | $r$ | $I_{11}$ | $I_{22}$ |
|---|---|---|---|---|---|---|---|---|
| Case 1 | 0.4 | 1 | 0.1 | 1 | 1 | 0.1 | 0.1 | 0.1 |
| Case 2 | 0.4 | 1 | 0.6 | 1 | 1 | 0.1 | 0.1 | 0.1 |
| Case 3 | 0.4 | 1 | 0.1 | 1 | 1 | 0.8 | 0.1 | 0.1 |

The sample size is $1\times10^7$ for numerical simulation. The PDFs obtained with short-time Gaussian approximation (STGA), path integration (PIS), equivalent linearization (EQL), and Monte-Carlo simulation (MIS) are represented by STGA, PIS, EQL, and MCS, respectively.

Eq. (10.11) presents the initial PDF distribution with $\mu_1=0$, $\mu_2=0$, $s_1=0.1$, and $s_2=0.1$

$$p(x^{(0)},v^{(0)})=\frac{1}{2\pi s_1 s_2}\exp\left[-\frac{(x^{(0)}-\mu_1)^2}{2s_1^2}-\frac{(v^{(0)}-\mu_2)^2}{2s_2^2}\right] \tag{10.11}$$

### 10.4.1 The Case of Slight Geometric Nonlinearity

This section considers that both geometric nonlinearity and excitation correlation are relatively slight. The state space is $[-4, 4]\times[-4, 4]$ with 80 uniform subintervals along the $x$ and $v$ directions. The short-time Gaussian approximation is adopted to obtain the transition PDFs (denoted by TPDFs). Monte-Carlo simulation is employed to verify its effectiveness. The time step is chosen as 0.4 in all cases. Figure 10.2 shows the TPDF results of Case 1 when the time step is 0.4. The initial values of $x$ and $v$ are $x_0=0$, $v_0=0$.

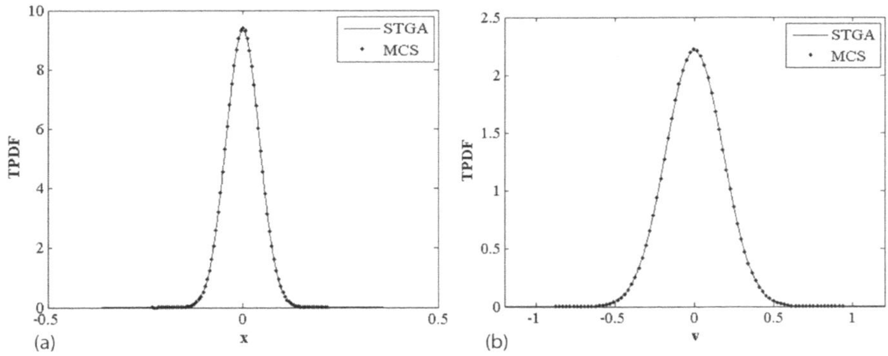


**FIGURE 10.2** Comparison of TPDFs of Case 1: (a) Displacement TPDFs; and (b) velocity TPDFs.

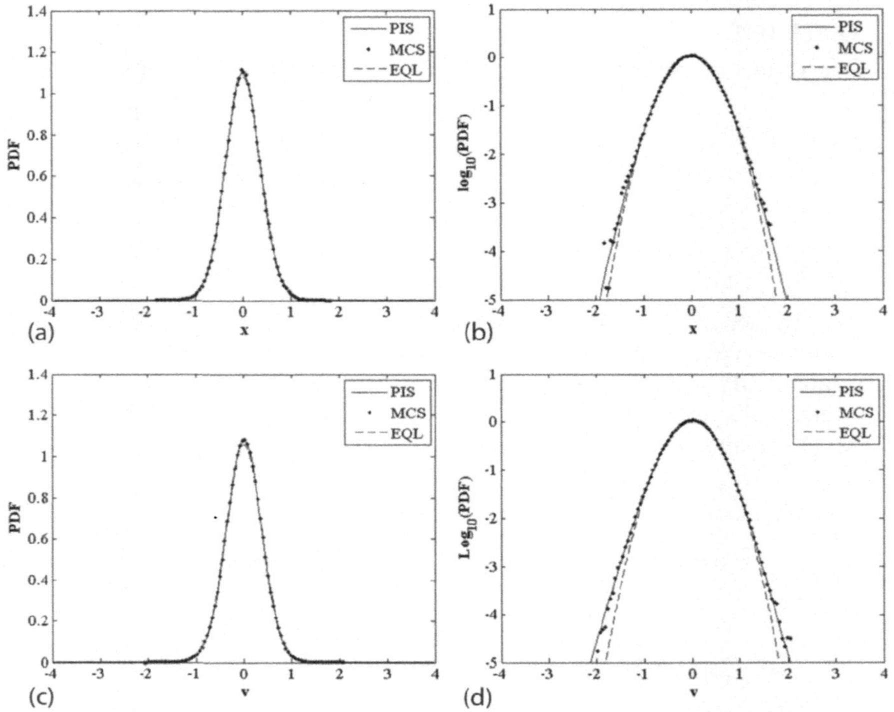


**FIGURE 10.3** PDF comparison in Case 1 (slight geometric nonlinearity): (a) Displacement PDFs; (b) logarithmic displacement PDFs; (c) velocity PDFs; and (d) logarithmic velocity PDFs.

From the implementation of solving the moment equations, the system becomes stationary at about $t=35$. Figure 10.3 presents a PDF comparison at $t=40$. Both Figures 10.3a and c present the roughly Gaussian PDFs for displacement and velocity. PIS presents a satisfactory agreement with MCS. For the tail PDF, the displacement PDF of MCS fluctuates because of the limited sample number. Furthermore, the tail PDFs have softening behaviors in the cases of both displacement and velocity.

Figures 10.4a–j show the evolution of joint probability density of the parametric marine riser system. On the whole, it shows that the joint PDF maintains a unimodal distribution. The peak value of joint probability density gradually decreases. The top view figures show that the tail region gradually expands until it reaches a stationary state. Figure 10.4k shows a stationary joint PDF acquired with the equivalent linearization method. Comparing Figures 10.4i and k, PIS is a little larger than EQL, and their distribution patterns are slightly different.

### 10.4.2 The Case of Strong Geometric Nonlinearity

This section considers strong nonlinearity in displacement. Similarly, the state space is also $[-4, 4]\times[-4, 4]$ with 80 uniform subintervals along the two directions. The time step is still 0.4, and the figures are not presented here.

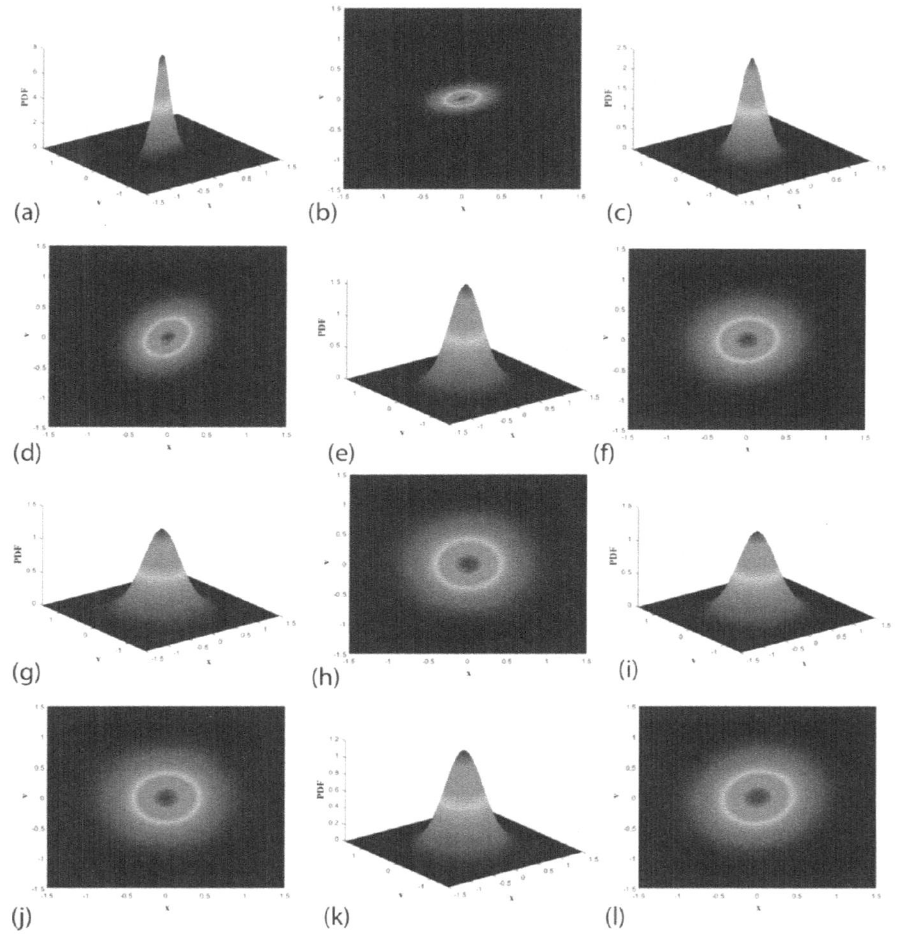

**FIGURE 10.4** Evolution of joint probability density in Case 1 (slight geometric nonlinearity): (a, b) $t=0.4$; (c, d) $t=2$; (e, f) $t=4$; (g, h) $t=16$; (i, j) $t=40$; (k, l) EQL.

The implementation of solving the moment equations shows that the PDF reaches its stationary state at about $t=35$. Figure 10.5 presents a PDF comparison at $t=40$. Figures 10.5a and c show that PIS coincides well with the simulation result (MCS), whereas EQL differs slightly from the simulation around the origin. The comparison proves that the path integration method is effective in this example. The tail PDF distribution of MCS is smoother than that in Case 1. Figures 10.5b and d show that PIS also works well, compared with MCS in the tail PDF region. However, displacement and velocity have opposite behaviors in the tail PDF regions. The former has a hardening distribution, whereas the latter has a softening distribution.

Figures 10.6a–j show the evolution of the joint probability density of the marine riser when the geometric nonlinearity is strong. In Figure 10.6a, the results on both sides of the v-axis are staggered and two peaks appear. The joint probability density

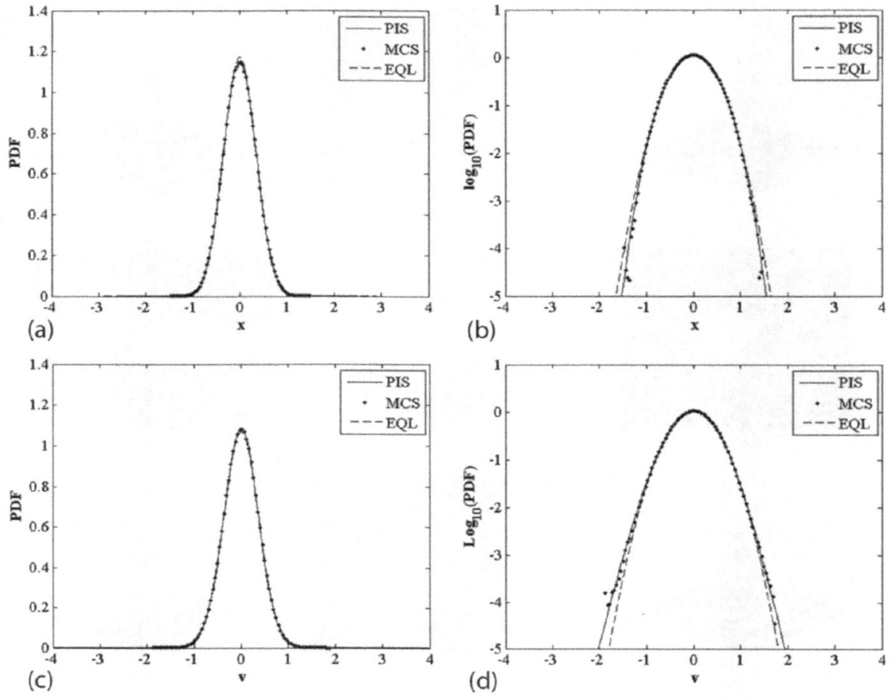


**FIGURE 10.5** PDF comparison in Case 2 (strong geometric nonlinearity): (a) Displacement PDFs; (b) logarithmic displacement PDFs; (c) velocity PDFs; (d) logarithmic velocity PDFs.

becomes and stays unimodel over time. The peak value gradually decreases and the tail region expands with time until reaching the stationary state. Figure 10.6k shows the stationary PDF with EQL. From the shape of the stationary PDF, there is not much difference between Figures 10.6i and k but there are differences in the peak value and the distribution range. Figure 10.6l shows that the joint probability density has a certain degree of inclination.

### 10.4.3 The Case of Strong Correlation between Excitations

In terms of the evolution of the moment equations, the system evolves into the stationary state at about $t=35$. Figure 10.7 shows a PDF comparison at $t=40$. Figures 10.7a and c show close agreement between PIS and MCS, whereas EQL differs significantly from the others. The tail of the displacement-simulated PDF fluctuates due to the limited sample number. The tail PDFs have softening behaviors in the cases of both displacement and velocity. The distribution of the PDFs of the system are not symmetrical in Figures 10.7b and d. Moreover, the asymmetry of the two is

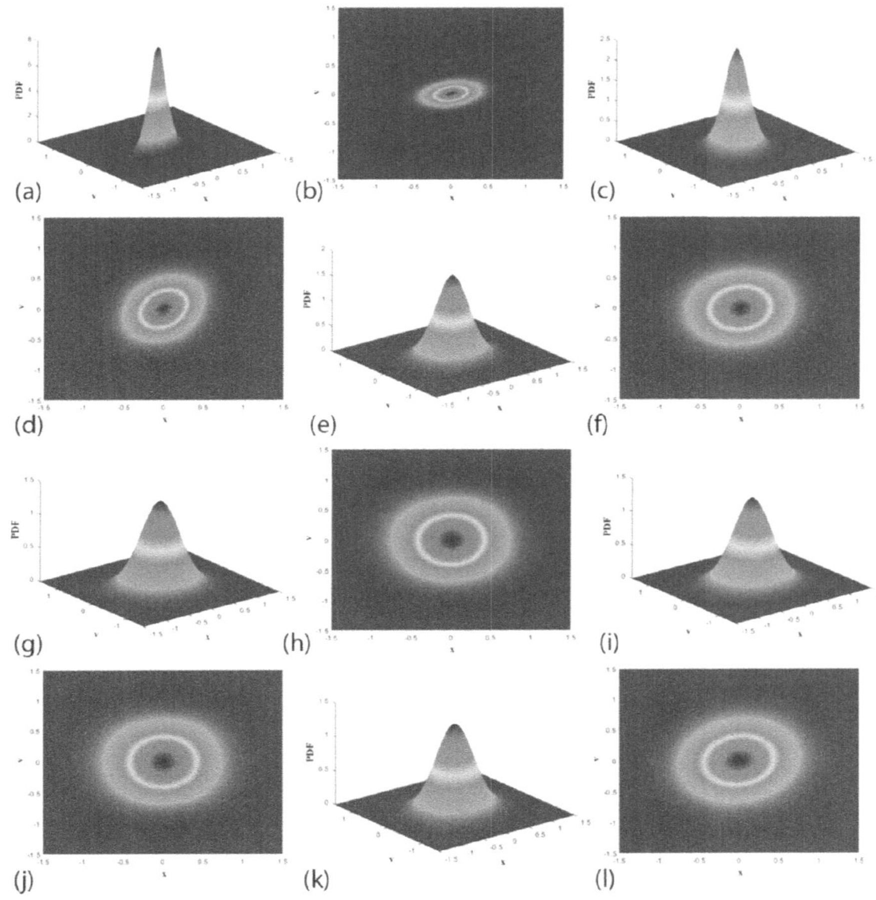


**FIGURE 10.6** Evolution of joint probability density in Case 2 (strong geometric nonlinearity): (a, b) $t=0.4$; (c, d) $t=2$; (e, f) $t=4$; (g, h) $t=16$; (i, j) $t=40$; (k,l) EQL.

opposite, with the displacement shifting to the positive axis, and the velocity shifting to the negative axis. Furthermore, velocity has a more significantly non-symmetrical distribution.

Figures 10.8a–j show the evolution of the joint probability density of the marine riser when the correlation between excitations is strong. In Figure 10.8b, the results on both sides of the v-axis are staggered. The joint probability density becomes and stays unimodal over time. The peak value gradually decreases and the tail region expands over time until reaching the stationary state. When the correlation between excitations increases, the joint probability density distribution is different from the first two cases, as shown in Figure 10.8d. In the positive and negative areas, the distribution is asymmetric, and the size of both ends is not the same. Figure 10.8k shows

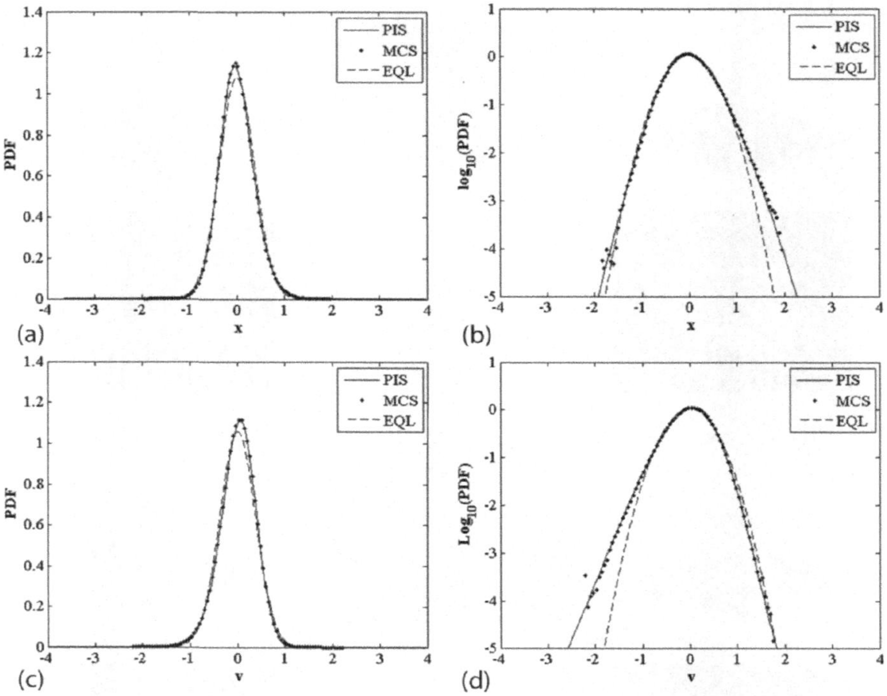


**FIGURE 10.7** PDF comparison in Case 3 (strong correlation between excitations): (a) Displacement PDFs; (b) logarithmic displacement PDFs; (c) velocity PDFs; (d) logarithmic velocity PDFs.

the stationary PDF with EQL. From the shape of the stationary PDF, Figure 10.8k shows that the distribution of the joint probability density has a significant tilt.

Figure 10.9 shows the PDFs obtained by PIS in three examples. When the geometric nonlinearity increases (Case 2), the displacement response decreases markedly, and the velocity response decreases slightly. The displacement PDF has a hardening tail distribution, whereas the velocity PDF has a softening tail distribution. When the correlation between two excitations is high (Case 3), the peaks of the PDFs increase slightly for displacement and velocity. The PDFs show softening behaviors both in displacement and velocity. From the figures of joint PDF, the joint PDF distribution shows a certain degree of rotation. The peak value of joint probability density decreases in the evolution process, and the tail region gradually expands until it reaches a stationary state.

## 10.5 CONCLUSION

In this chapter, the response evolution of a marine riser is solved by path integration. First, the random vibration of the marine riser is studied by analyzing the

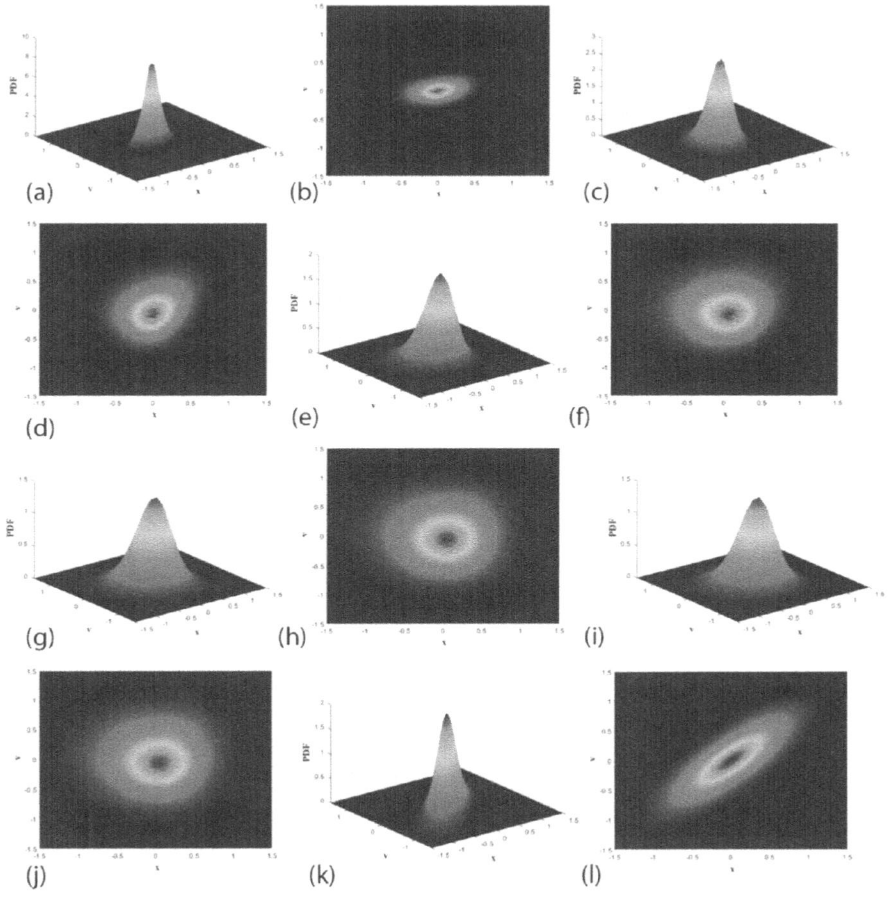

**FIGURE 10.8** Evolution of joint probability density in Case 3 (strong correlation between excitations): (a, b) t=0.4; (c, d) t=2; (e, f) t=4; (g, h) t=16; (i, j) t=40; (k, l) EQL.

environment around the riser, and its equation of motion is established by the Galerkin method, leading to a differential equation. Secondly, based on the differential equation, the path integration method is employed to study the response PDF evolution. In the path integration procedure, the transition PDF is approximated by short-time Gaussian PDF. The integration is conducted with the Gauss-Legendre scheme. A further parametric analysis is conducted on the effects of different parameters for the marine riser. The PDF evolution of the response is discussed in detail for the marine riser. When geometric nonlinearity is slight, the PDFs are roughly Gaussian, but both displacement and velocity have softening behaviors in their tail PDF region compared with the PDF distribution of EQL. When geometric nonlinearity becomes strong, displacement has a hardening PDF distribution, whereas velocity has a softening PDF distribution. When the correlation between excitations is strong,

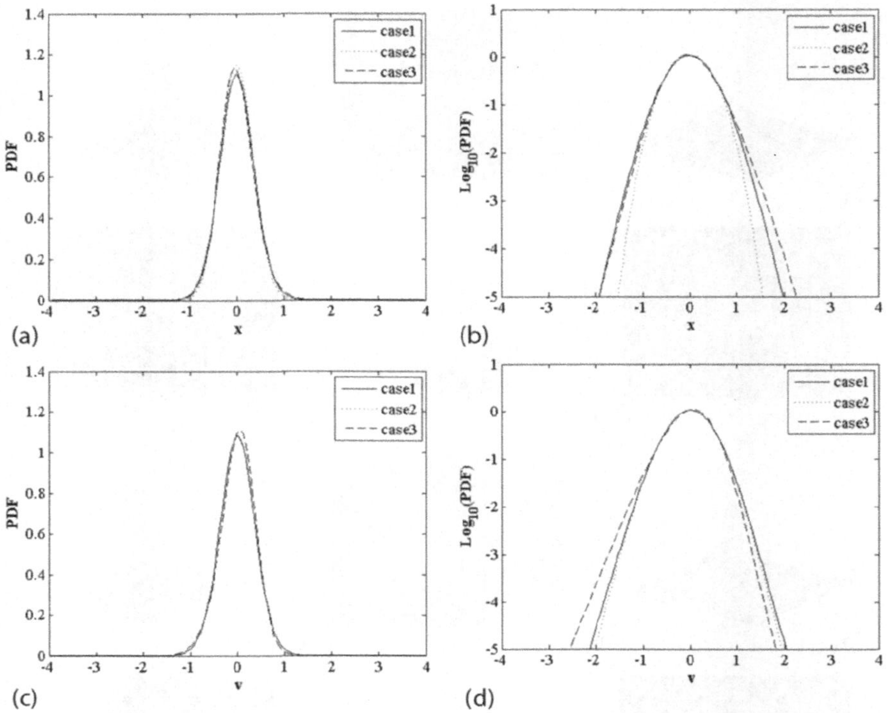


**FIGURE 10.9** Comparison of the PIS results in the three cases: (a) Displacement PDFs; (b) logarithmic displacement PDFs; (c) velocity PDFs; (d) logarithmic velocity PDFs.

both displacement and velocity have softening PDF distributions. Furthermore, their PDF distributions are not symmetrical, and velocity has a more significantly non-symmetrical distribution.

## REFERENCES

Alevras P., and D. Yurchenko. 2016. GPU computing for accelerating the numerical path integration approach. *Computers & Structures* 171:46–53.

Alfosail F.K., and M.I. Younis. 2019. Three-to-one internal resonance of inclined marine riser. *International Journal of Non-Linear Mechanics* 109:107–117.

Barvinsky A.O. 1998. Solution of quantum Dirac constraints via path integral. *Nuclear Physics B* 520(1–2):533–560.

Cabrera-Miranda J.M., and J.K. Paik. 2017. On the probabilistic distribution of loads on a marine riser. *Ocean Engineering* 134:105–118.

Chai W., L. Dostal, A. Naess, and B.J. Leira. 2017. Comparative study of the path integration method and the stochastic averaging method for nonlinear roll motion in random beam seas. *Procedia Engineering* 199:1110–1121.

Gaidai O., A. Naess, and M. Dimentberg. 2017. Response statistics of rotating shaft with non-linear elastic restoring forces by path integration. *Journal of Sound and Vibration* 400:113–121.

Gaidai O., P. Dou, A. Naess, M. Dimentberg, Y. Cheng, and R. Ye. 2019. Nonlinear 6D response statistics of a rotating shaft subjected to colored noise by path integration on GPU. *International Journal of Non-Linear Mechanics* 111:142–148.

Grosjean C.C., and M.J. Goovaerts. 1973. A perturbational calculation relative to the energy states of the linear harmonic oscillator in the path-integral formalism of nonrelativistic quantum mechanics. *Physica* 70(2):243–256.

Han Q., W. Xu, and J.Q. Sun. 2016. Stochastic response and bifurcation of periodically driven nonlinear oscillatoes by the generalized cell mapping method. *Physica A* 458(15):115–125.

Hasnijeh S.G., M. Poursina, B.J. Leira, H. Karimpour, and W. Chai. 2019. Stochastic dynamics of a nonlinear time-varying spur gear model using an adaptive time-stepping path integration method. *Journal of Sound and Vibration* 447:170–185.

He J.W., and Y.M. Low. 2012. An approach for estimating the probability of collision between marine risers. *Applied Ocean Research* 35:68–76.

Narayanan S., and P. Kumar. 2012. Numerical solutions of Fokker-Planck equation of nonlinear systems subjected to random and harmonic excitations. *Probabilistic Engineering Mechanics* 27(1):35–46.

Ni P., J. Li, H. Hao, and Y. Xia (2018). Stochastic dynamic analysis of marine risers considering Gaussian system uncertainties. *Journal of Sound and Vibration* 416:224–243.

Ni P., J. Li, H. Hao, Y. Xia, and X. Du (2019). Stochastic dynamic analysis of marine risers considering fluid-structure interaction and system uncertainties. *Engineering Structures* 198:109507.

Patel M.H., and H.I. Park. 1995. Combined axial and lateral responses of tensioned buoyant platform tethers. *Engineering Structure* 17(10):687–695.

Ren Z., and W. Xu. 2020. An improved path integration method for nonlinear systems under Poisson white noise excitation. *Applied Mathematics and Computation* 373:125036.

Thorsen M.J., N.R. Challabotla, S. Saevik, and O.J. Nydal. 2019. A numerical study on vortex-induced vibrations and the effect of slurry density variations on fatigue of ocean mining risers. *Ocean Engineering* 174:1–13.

Xie W. X., W. Xu, and L. Cai. 2006. Study of the Duffing-Rayleigh oscillator subject to harmonic and stochastic excitations by path integration. *Applied Mathematics and Computation* 172(6):1212–1224.

Yang C., J. Du, Z. Cheng, Y. Wu, and C. Li. 2020. Flexibility investigation of a marine riser system based on an accurate and efficient modelling and flexible multibody dynamics. *Ocean Engineering* 207:107407.

Yin D.C., E. Passano, H. Lie, G. Grytoyr, K. Aronsen, M. Tognarelli, and E.B. Kebadze. 2019. Experimental and numerical study of a top tensioned riser subjected to vessel motion. *Ocean Engineering* 171(1):565–574.

Yu J.S., G.Q. Cai, and Y.K. Lin. 1997. A new path integration procedure based on Gauss-Legendre scheme. *International Journal of Non-Linear Mechanics* 32(4):759–768.

Zhang J., and Y.G. Tang. 2014. Mathieu instability analysis of deepwater top-tensioned risers. *Journal of Ship Mechanics* 18(9):1142–1150.

Zhu H. T. 2017. Non-stationary response of a van der Pol-Duffing oscillator under Gaussian white noise. *Meccanica* 52(4–5):833–847.

Zhu H.T., and L.L. Duan. 2016. Probabilistic solution of non-linear random ship roll motion by path integration. *International Journal of Non-Linear Mechanics* 83:1–8.

# 11 Solution of System of PDE Governed in Natural Convective Flow in a Rectangular Porous Cavity

*P. Alam and S. Kapoor*

## CONTENTS



## 11.1 INTRODUCTION

The mathematical solution of systems by partial differential equations (PDE) is an interesting area used by many researchers. Training in advanced mathematical techniques is an essential requirement for young researchers and engineers. The idea of heat transport through permeable media has attracted extensive attention over a long period because of its wide range of uses in design science and the applied sciences, from toxin transport paper fabrication (Koponen et al. 1998), and geophysics and oil platform design (King et al. 1999), to marine life science the extraction of geothermal energy supplies, design of low-temperature protection gear, packed-bed synergist reactors, heat stockpiling beds, and atomic waste removal, etc. The vast majority of the work has been centered around isotropic and homogeneous permeable media. Understanding the dynamic movement of liquid through a permeable medium as a result of common convection is well recorded in the survey work of Nield and Bejan (2006) and Ingham and Pop (1998). In the current chapter, we are dealing with free convection in a rectangular hole filled with a permeable medium, the four dividers of which have distinctive fractional warming or cooling characteristics.

A detailed review of the literature on the above-characterized second class of regular convection in this chapter shows that, under liquid conditions, a reasonable number of papers are available. For instance, an exploratory and mathematical investigation of free convective heat transfer in this cavity, described by a discrete warming at the lower divider and cooling from the vertical dividers, was analyzed. A mathematical examination of typical convection in air in a vertical square pit with limited isothermal warming from underneath but uniform cooling from the sidewalls was explored. A similar issue, using a steady-state heat source rather than the limited isothermal heat source at the base divider, was analyzed by Sharif and Mohammad (2005). They researched the impact of viewpoint proportion and the tendency of the hole toward heat transfer. Normal convection in a square walled-in area, warmed occasionally from part of the base divider, has been researched by Lakhal et al. (1995).

The impact of warmer and cooler areas on common convection in square holes has been investigated by Turgoklu and Yucel (1995). Normal convection in rectangular tanks warmed locally from below has been defined mathematically by Sarris et al. (2004), who found that, for low Rayleigh (*Ra*) numbers, the movement of heat is dominated by conduction, whereas, at higher *Ra* values, convection becomes dominant. The increment of the tank angle proportion and the width of the heated strip increases the liquid stream and raises the temperature of the liquid. This makes the glass-dissolve more homogeneous, improving the eventual outcome. Recently, normal convection in an air-filled 2D square walled-in area, warmed by a consistent source from beneath and cooled from above, was defined mathematically by Nader et al. (2007). They considered an assortment of heat-limited conditions at the top and sidewalls. Modeling was performed for two sizes of the heat source, i.e., a small and a very large source, representing 20% and 80% of the complete length of the base divider, respectively. Their outcomes were presented as smoothed-out and isothermal plots in terms of the variation in the Nusselt number and the most extreme temperature at the heat source surface. Additionally, they have revealed correlations among the various heating designs.

Various specialists have also completed exploratory and mathematical examinations to contemplate liquid stream function and heat transfer rate in discretely warmed permeable spaces. mathematically researched the impacts of stratification on heat convection in a flat pit loaded with a liquid-immersed permeable medium with a limited heat source at the base surface and directly changing temperature at the side dividers. Robillard (1988) considered numerous consistent states in a connected permeable medium with confined heating from underneath. Lai et al. (1990) and Lai and Kulacki (1991) mathematically examined free and blended convection in flat permeable layers with numerous, isothermal, discrete hotspots for different Rayleigh and Peclet numbers. Hsiao et al. (1994) investigated natural convection in an inclined porous cavity with discrete heat sources on a wall. The effects of variable heat source spacing and heat source width on heat transfer increases and pressure drops in partially porous channels with discrete heat sources were investigated. Saeid and Pop (2005) studied natural convection mathematically in a square cavity with a discrete heat source on a vertical wall with isothermal and isoflux boundary conditions. More recently, a numerical study of double-diffusive convective flow of a binary mixture in a porous medium subject to localized heating and salting from one side was performed by Zhao et al. (2008). In a continuation of the above study, Natesan (2017) focused on the mathematical solution of such types of problems by including the properties of a nanofluid. Recently, Hassinet (2019) reported the mathematical solution of a cavity flow problem, using the finite volume technique, with the Darcy-Brinkman model for the momentum conservation equation, which allows for the no-slip boundary condition on a solid wall. Hassinet (2019) found that, as the permeability of the porous medium decreased, the temperature and concentration contours became more parallel to the vertical walls, indicating the tendency toward a quasi-conductive regime. The main effects of decreasing the Darcy number are predicted to below retardation effect and a suppression of the overall heat and mass transfer in the enclosed space.

In the current chapter, a mathematical examination of the characteristic convection in a square cavity is introduced. The stream is instigated by steady fractional warming at the left vertical divider and partial cooling at the correct vertical divider alongside the adiabatic rest dividers. In this examination, consideration is given to understanding the impact of medium permeability through the Darcy number ($Da$) and the heat power source, i.e., the Rayleigh number, $Ra$, on the liquid stream design, just as on the local and average heat transfer rates. The outcomes are introduced with respect to stream function $\psi$, temperature $\theta$, and Nusselt numbers (local Nusselt number, $NuL$, and average Nusselt number, $Nu$).

## 11.2 MODEL FORMULATION

Consider a two-dimensional liquid-immersed permeable medium encased in a square hole of length L, as indicated schematically in Figure 11.1. The stream is activated by halfway warming of the left vertical wall and incomplete cooling at the right vertical divider alongside adiabatic level dividers. The Boussinesq equations are adapted for the liquid properties to relate thickness changes to temperature changes, coupling the temperature field to the stream field in this way. The accompanying assumptions are made here for permeable media:

- The media just as thermal diffusivity are thought to be isotropic.
- The actual circumstance is depicted by local harmony model conditions.
- The thermo-actual properties of the liquid are thought to be constant except for the density depend on the body force term in the momentum equation, which is fulfilled by the Boussinesq estimate.
- The internal radiation in the porous medium is ignored due to the low absolute temperature in the cavity.

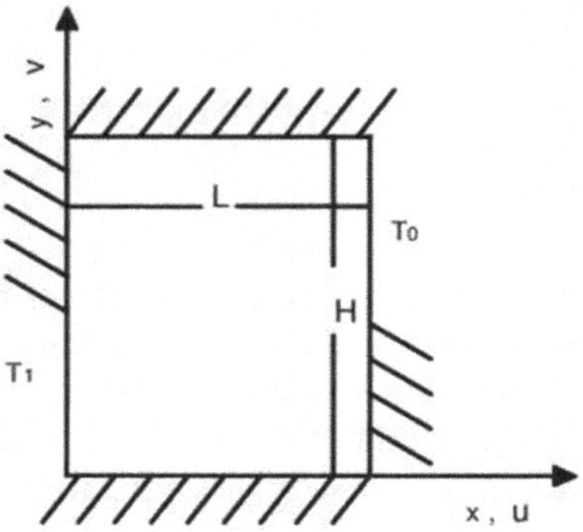


**FIGURE 11.1** Schematic diagram.

In expressing the equations for flow in the permeable medium, it should be noted that the Darcy model is an empirical formula relating the pressure gradient, the gravitational force, and the bulk viscous resistance in low-permeability media. The Brinkman-extended term is needed and included to satisfy a no-slip boundary condition at the solid walls, whereas the Forchheimer drag term is included in view of the form drag. The material derivative is used to extend the Darcy model. The Brinkman-extended non-Darcy model, including material derivatives, has been used in the momentum equations. The system of equations in the Cartesian coordinate system ($x$, $y$) for the steady-state flow, as well as for heat transfer, is given by the governing equations (Section 11.3).

## 11.3 GOVERNING EQUATIONS

$$\frac{\partial u}{\partial x}+\frac{\partial v}{\partial y}=0 \tag{11.1}$$

$$u\frac{\partial u}{\partial x}+v\frac{\partial u}{\partial y}=-\frac{1}{\rho}\left(\frac{\partial p}{\partial x}\right)+\nu\left(\frac{\partial^2 u}{\partial x^2}+\frac{\partial^2 u}{\partial y^2}\right)-\frac{\nu}{K}u \tag{11.2}$$

$$u\frac{\partial v}{\partial x}+v\frac{\partial v}{\partial y}=-\frac{1}{\rho}\left(\frac{\partial p}{\partial x}\right)+\nu\left(\frac{\partial^2 v}{\partial x^2}+\frac{\partial^2 v}{\partial y^2}\right)-\frac{\nu}{K}v+g\beta_T\left(T-T_0\right) \tag{11.3}$$

$$u\frac{\partial T}{\partial x}+v\frac{\partial T}{\partial y}=\alpha_e\left(\frac{\partial^2 T}{\partial x^2}+\frac{\partial^2 T}{\partial y^2}\right) \tag{11.4}$$

with the boundary conditions

$$u(x,0)=u(x,L)=u(0,y)=u(L,y)=0$$

$$v(x,0) = v(x,L) = v(0,y) = v(L,y) = 0$$

$$\frac{\partial T}{\partial y}(x,0) = 0, \ \frac{\partial T}{\partial y}(x,L) = 0$$

$$T(0,y) = T_1; \ 0 \le y \le \frac{L}{2}$$

$$\frac{\partial T}{\partial x} = 0; \ \frac{L}{2} \le y \le L$$

$$T(L,y) = T_0; \ \frac{L}{2} \le y \le L$$

$$\frac{\partial T}{\partial x} = 0; \ 0 \le y \le \frac{L}{2}$$

In the above equations, $L$ is the length of the cavity, and $V(u,v)$, $p, T, \alpha_e$ and $\rho$ are the flow velocity, pressure, temperature, effective thermal diffusivity, and fluid density, respectively. Using non-dimensional quantities:

$$X = \frac{x}{L}, Y = \frac{y}{L}, U = \frac{Lu}{\alpha_e}, V = \frac{Lv}{\alpha_e}, P = \frac{pL^2}{\rho\alpha_e^2}, \text{and } \theta = \frac{(T - T_0)}{(T_1 - T_0)}$$

the above system of differential equations may be written as non-dimensional equations (Section 11.4).

## 11.4 NON-DIMENSIONAL EQUATIONS

$$\frac{\partial U}{\partial X} + \frac{\partial V}{\partial Y} = 0 \tag{11.5}$$

$$U\frac{\partial U}{\partial X} + V\frac{\partial U}{\partial Y} = -\frac{\partial P}{\partial X} + Pr\left[\nabla^2 U\right] - \frac{Pr}{Da}U \tag{11.6}$$

$$U\frac{\partial V}{\partial X} + V\frac{\partial V}{\partial Y} = -\frac{\partial P}{\partial Y} + Pr\left[\nabla^2 V\right] - \frac{Pr}{Da}V + Ra\,Pr\theta \tag{11.7}$$

$$U\frac{\partial \theta}{\partial X} + V\frac{\partial \theta}{\partial Y} = \frac{\partial^2 \theta}{\partial X^2} + \frac{\partial^2 \theta}{\partial Y^2} \tag{11.8}$$

where $\nabla^2 = \frac{\partial^2}{\partial x^2} + \frac{\partial^2}{\partial y^2}$

with the boundary conditions

$$U(X,0)=U(X,1)=U(0,Y)=U(1,Y)=0$$

$$V(X,0)=V(X,1)=V(0,Y)=V(1,Y)=0$$

$$\frac{\partial\theta}{\partial Y}(X,0)=\frac{\partial\theta}{\partial Y}(X,1)=0$$

$$\theta(0,Y)=1;\ 0\le Y\le\frac{1}{2}$$

$$\frac{\partial\theta}{\partial X}=0;\ \frac{1}{2}\le Y\le 1$$

$$\theta(1,Y)=0;\ \frac{1}{2}\le Y\le 1$$

$$\frac{\partial\theta}{\partial X}=0;\ 0\le Y\le\frac{1}{2}$$

where *U*, *V*, $\theta$, and *P* are dimensionless velocity components in the *x*-direction, dimensionless velocity components in the *y*-direction, dimensionless temperature, and dimensionless pressure, respectively. In the above equation, *Ra*, *Da*, and *Pr* are the Rayleigh number, Darcy number, and Prandtl number, respectively, and are defined as

$Da$ = Darcy number = $K/L^2$
$Pr$ = Prandtl number = $\nu/\alpha_e$
$Ra$ = Rayleigh number = $g\beta L^3\,(T2 - T1)/\alpha_e\nu$

## 11.5 SOLUTION PROCEDURE

The arrangement of differential conditions has been tackled by using the Galerkin finite element method (FEM). The coherence condition, Eq. (11.5), will be used as a requirement to fulfill the mass preservation, and this requirement might be used to determine the weight circulation as cited by specialists. To solve Eqs. (11.5)–(11.8), the penalty FEM is used in which the pressure *P* is eliminated by a penalty parameter $\gamma$ and the incompressibility criterion given by Eq. (11.5), which results in:

$$P=-\gamma\left\{\frac{\partial U}{\partial X}+\frac{\partial V}{\partial Y}\right\} \tag{11.9}$$

$$U\frac{\partial U}{\partial X}+V\frac{\partial U}{\partial Y}=\gamma\left\{\frac{\partial U}{\partial X}+\frac{\partial V}{\partial Y}\right\}+Pr\left\{\frac{\partial^2 U}{\partial X^2}+\frac{\partial^2 U}{\partial Y^2}\right\}-\frac{Pr}{Da}U \tag{11.10}$$

$$U\frac{\partial V}{\partial X}+V\frac{\partial V}{\partial Y}=\gamma\left\{\frac{\partial U}{\partial X}+\frac{\partial V}{\partial Y}\right\}$$
$$+Pr\left\{\frac{\partial^2 V}{\partial X^2}+\frac{\partial^2 V}{\partial Y^2}\right\}-\frac{Pr}{Da}V+PrRa\theta \tag{11.11}$$

The unknown variables, velocity ($U$, $V$) and temperature, $\theta$, are approximated to as shown below

$$U=\sum_{k=1}^{N}U_k\Phi_k(X,Y),\ V=\sum_{k=1}^{N}V_k\Phi_k(X,Y)$$
$$\text{And}\quad \theta=\sum_{k=1}^{N}U_k\theta_k(X,Y) \tag{11.12}$$

The (nonlinear) residual corresponding to the given equations are:

$$R_i^1=\sum_{k=1}^{N}U_k\int_{\Omega}^{A}\left[\left(\sum_{k=1}^{N}U_k\Phi_k\right)\frac{\partial\Phi_k}{\partial X}+\left(\sum_{k=1}^{N}V_k\Phi_k\right)\frac{\partial\Phi_k}{\partial Y}\right]\Phi_i dXdY$$
$$+\gamma\left[\sum_{k=1}^{N}U_k\int_{\Omega}^{A}\frac{\partial\Phi_k}{\partial X}\frac{\partial\Phi_i}{\partial X}dXdY+\sum_{k=1}^{N}V_k\int_{\Omega}^{A}\frac{\partial\Phi_i}{\partial X}\frac{\partial\Phi_k}{\partial Y}dXdY+\right] \tag{11.13}$$
$$+Pr\sum_{k=1}^{N}U_k\int_{\Omega}^{A}\left[\frac{\partial\Phi_k}{\partial X}\frac{\partial\Phi_i}{\partial X}+\frac{\partial\Phi_k}{\partial Y}\frac{\partial\Phi_i}{\partial Y}\right]dXdY+\frac{Pr}{Da}\int_{\Omega}^{A}\left(\sum_{k=1}^{N}U_k\Phi_k\right)\Phi_i dXdY$$

$$R_i^2=\sum_{k=1}^{N}V_k\int_{\Omega}^{A}\left[\left(\sum_{k=1}^{N}U_k\Phi_k\right)\frac{\partial\Phi_k}{\partial X}+\left(\sum_{k=1}^{N}V_k\Phi_k\right)\frac{\partial\Phi_k}{\partial Y}\right]\Phi_i dXdY$$
$$+\gamma\left[\sum_{k=1}^{N}U_k\int_{\Omega}^{A}\frac{\partial\Phi_i}{\partial Y}\frac{\partial\Phi_k}{\partial X}dXdY+\sum_{k=1}^{N}V_k\int_{\Omega}^{A}\frac{\partial\Phi_k}{\partial Y}\frac{\partial\Phi_i}{\partial Y}dXdY\right]$$
$$+Pr\sum_{k=1}^{N}V_k\int_{\Omega}^{A}\left[\frac{\partial\Phi_k}{\partial X}\frac{\partial\Phi_i}{\partial X}+\frac{\partial\Phi_k}{\partial Y}\frac{\partial\Phi_i}{\partial Y}\right]dXdY+\frac{Pr}{Da}\int_{\Omega}^{A}\left(\sum_{k=1}^{N}U_k\Phi_k\right)\Phi_i dXdY \tag{11.14}$$
$$-RaPr\sum_{k=1}^{N}V_k\int_{\Omega}^{A}\left(\sum_{k=1}^{N}\theta_k\Phi_k\right)\Phi_i dXdY$$

and

$$R_i^3 = \sum_{k=1}^{N} \theta_k \int_{\Omega}^{A} \left[ \left( \sum_{k=1}^{N} U_k \Phi_k \right) \frac{\partial \Phi_k}{\partial X} + \left( \sum_{k=1}^{N} V_k \Phi_k \right) \frac{\partial \Phi_k}{\partial Y} \right] \Phi_i dXdY$$
$$+ \sum_{k=1}^{N} \theta_k \int_{\Omega}^{A} \left[ \frac{\partial \Phi_i}{\partial X} \frac{\partial \Phi_k}{\partial X} + \frac{\partial \Phi_i}{\partial Y} \frac{\partial \Phi_k}{\partial Y} \right] \tag{11.15}$$

In this method, the bi-quadratic premise capacities are used for approximations to the obscure field factors and their integration obtained by the three-point Gaussian formula under the prevailing conditions. The residuals obtained might be re-written in the accompanying grid structure

$$\left[ A_1 + \gamma A_2 \right] z = F \tag{11.16}$$

where $A_1$ and $A_2$ are coefficient matrices obtained from the Jacobian matrix of the residuals, and $z$ and $F$ are the unknown and known vectors, respectively. The estimated value of the penalty boundary (γ) is taken to be 108. The Newton-Raphson iterative technique has been used to tackle the nonlinear residual (Eqs. (11.16)–(11.18)). At every iteration, the linear system of equations of the order (3N × 3N) is

$$J\left(a^n\right)\left[a^n - a^{n+1}\right] = R\left(a^n\right) \tag{11.17}$$

where $n$ is the iterative index and $J$ the Jacobian matrix. The iterative procedure is stopped when the convergence criterion is satisfied. The convergence criterion is

$$\left[ \Sigma \left( R_i^{(j)} \right)^2 \right]^{\frac{1}{2}} \leq 10^{-5} \tag{11.18}$$

A nine-node bi-quadratic element, with each element in the $(X, Y)$ plane mapped to a unit square domain in the $(\xi, \eta)$ plane, using iso-parametric mapping. The domain integrals in the residual equations are evaluated using nine node bi-quadratic basis functions in $(\xi, \eta)$ domain as:

$$X = \sum_{i=1}^{9} X_i \Phi_i(\xi, \eta) \quad \text{and} \quad Y = \sum_{i=1}^{9} Y_i \Phi_i(\xi, \eta) \tag{11.19}$$

where $\Phi i$ $(\xi, \eta)$ are the local bi-quadratic basis functions on the $(\xi, \eta)$ domain. The integrals in Eqs. (11.16) and (11.18) can be evaluated in $(\xi, \eta)$ domains, using the following transformation:

$$\begin{bmatrix} \dfrac{\partial \Phi_i}{\partial X} \\ \dfrac{\partial \Phi_j}{\partial Y} \end{bmatrix} = \frac{1}{J} \begin{bmatrix} \dfrac{\partial Y}{\partial \eta} & -\dfrac{\partial Y}{\partial \xi} \\ \dfrac{\partial X}{\partial \eta} & -\dfrac{\partial X}{\partial \xi} \end{bmatrix} \begin{bmatrix} \dfrac{\partial \Phi_i}{\partial \xi} \\ \dfrac{\partial \Phi_j}{\partial \eta} \end{bmatrix}$$

and

$$dXdY = Jd\xi d\eta \tag{11.20}$$

where

$$J = \frac{\partial (X,Y)}{\partial (\xi,\eta)} = \begin{bmatrix} \dfrac{\partial X}{\partial \xi} & \dfrac{\partial X}{\partial \eta} \\ \dfrac{\partial Y}{\partial \xi} & \dfrac{\partial Y}{\partial \eta} \end{bmatrix}$$

## 11.6 STREAM FUNCTION AND NUSSELT NUMBER

The stream function $\psi$ for two-dimensional flows is obtained from the velocity components $U$ and $V$. The relationships between the stream function, $\psi$, and the velocity components are:

$$U = \frac{\partial \psi}{\partial Y} \text{ and } V = -\frac{\partial \psi}{\partial X} \tag{11.21}$$

From the above equation, we can obtain the following relationship:

$$\frac{\partial^2 \psi}{\partial^2 X} + \frac{\partial^2 \psi}{\partial^2 Y} = \frac{\partial U}{\partial Y} - \frac{\partial V}{\partial X} \tag{11.22}$$

The stream function $\psi$ is approximated to using the basis function set ($\Phi$) as $\psi = \sum_{k=1}^{9} \psi_k \Phi_k (X,Y)$ and the residual equation corresponding to (11.22) is

$$\begin{aligned} R_i^4 = {} & \sum_{k=1}^{N} \psi_k \int_{\Omega} \left[ \frac{\partial \Phi_k}{\partial X}\frac{\partial \Phi_i}{\partial X} + \frac{\partial \Phi_k}{\partial Y}\frac{\partial \Phi_i}{\partial Y} \right] dXdY \\ & + \sum_{k=1}^{N} U_k \int_{\Omega} \Phi_i \frac{\partial \Phi_k}{\partial Y} dXdY - \sum_{k=1}^{N} V_k \int_{\Omega} \Phi_i \frac{\partial \Phi_k}{\partial X} dXdY \end{aligned} \tag{11.23}$$

Since the no-slip boundary condition is valid at all boundaries, $\psi$ will be zero at all grid points on the boundaries of the cavity. The heat transfer coefficient in terms of the local Nusselt number is obtained as

$$Nu = -\frac{\partial \theta}{\partial n} \tag{11.24}$$

where $n$ denotes the typical direction of the plane. The normal derivative is obtained by the biquadratic basis functions in the $(\xi, \eta)$ domain with the help of equations (11.25) and (11.26). The local Nusselt number at the left side wall ($Nu_L$) is defined as:

$$Nu_L = -\sum_{i=1}^{9} \theta_i \frac{\partial \theta_i}{\partial X} \tag{11.25}$$

The average heat transfer rate (Nusslet number, $Nu$) at the left side wall is:

$$Nu = \frac{1}{A} \int_{0}^{A/2} Nu_L dY \tag{11.26}$$

## 11.7 INTERPRETATION OF RESULTS

In this section, we try to shine a light on a rigorous study with mathematical methods, made to solve the system of PDE and to observe the dependence of stream function, temperature and heat transfer rate (Nusselt number) on individual physical parameters present in the system. The heat source intensity in the square cavity filled with a permeable medium is governed by the Rayleigh number ($Ra$), whereas the medium permeability is determined by the Darcy number ($Da$). Based on the non-dimensional analysis, as well as the real data available in the literature (Nield and Bejan 2006), a wide range of different parameters ($10^{-6} \leq Da \leq 10^{-1}$) and ($10 \leq Ra \leq 10^{6}$) are obtained here. The main emphasis is given to the computational domain, which is divided into $30 \times 30$ bi-quadratic elements and $51 \times 51$ grid points are taken in each element. In order to establish the method successfully, comparison with the published results of Lauriat and Prasad (1989) is shown under some special conditions, in which the Brinkman-extended non-Darcy model is used without the Forchheimer term; we have also incorporated some changes by changing the boundary conditions appropriately and found a very close agreement with our mathematical results with respect to published results, as shown in Table 11.1.

**TABLE 11.1**
**Comparison with Published Results**

| *Da* | *Ra* | Lauriat and Prasad (1989) (Brinkman-extended) | This chapter |
|---|---|---|---|
| $10^{-4}$ | $10^{5}$ | 1.07 | 1.0752 |
| $10^{-5}$ | $10^{6}$ | 3.09 | 3.0664 |
| $10^{-6}$ | $10^{7}$ | 1.07 | 1.0755 |

In the complete computation study, we have fixed $Pr$ at 100; this is under an assumption that the fluid might be oil. The other emphasis is given with regard to the influence of medium permeability *via* the Darcy number ($Da$) and heat source intensity, i.e., the Rayleigh number ($Ra$), on flow configuration, as well as on the heat transfer mechanism.

In the first part of the study, permeability is a measure of the flow strength of the medium, and also acts as a measure of the conductivity of the fluid flows. In general, high permeability produces a strong flow, whereas low permeability produces a weak flow. In this study, the effect of medium permeability is considered by studying the effect of the Darcy number ($Da$) on the flow configuration (stream function as well as temperature) as well as on heat transfer rates in terms of the Nusselt number (local Nusselt number, $Nu_L$, as well as average Nusselt number, $Nu$). To understand the impact of Darcy number ($Da$) on the stream function, as well as temperature, a comparative study is made for $Ra$ equal to $10^5$ and $10^6$.

As can be seen from Figures 11.2a–b and Figures 11.3a–b, initially, the change in the maximum value of the stream function is negligible for $Da = 10^{-7}$ to $10^{-5}$, whereas a significant change occurs at $Da = 10^{-4}$. This shows that the strength of the flow inside the cavity increases around 10 times at $Da = 10^{-5}$ and at $Da = 10^{-4}$. The corresponding temperatures for different values of $Da$ are plotted in Figures 11.2a–b at $Ra = 10^5$.

It can be seen from Figures 11.3a–b that the temperature is separated vertically because of the incomplete warming at the left half vertical divider and partial cooling at the upper right half divider. The warmed part of the liquid moves higher than the rest of the liquid in the pit, moving away from the warmed part of the hole. This shows that the strength of the flow inside the cavity increases around 10 times for different values of $Da$.

Figures 11.4a–b and Figures 11.5a–b illustrate the stream function and isotherm contours of the numerical results for various $Da = 10^{-7}$ to $10^{-4}$ at $Ra = 10^6$. In general, the fluid circulation is strongly dependent on the Darcy number, as can be seen from Figures 11.4a–b and Figures 11.5a–b. From these figures, it can be seen that the flow is very weak when the Darcy number varies from $10^{-7}$ to $10^{-6}$, whereas the flow becomes stronger at $Da = 10^{-5}$ and $10^{-4}$. The variation in the maximum values of stream function ($\psi_{max}$) is negligible up to $Da = 10^{-6}$ and it increases by approximately ten times at $Da = 10^{-6}$ and $Da = 10^{-5}$; again, if we increase $Da$ by one order of magnitude (i.e., to $Da = 10^{-4}$), the value of $\psi_{max}$ increases by one order of $\psi_{max}$ at $Da = 10^{-5}$.

From Figure 11.5, it tends to be seen that the temperature is defined vertically because of the partial heating at the left half vertical divider and partial cooling at the upper right half divider. The warmed portion of the liquid rises higher than the rest of the liquid in the cavity and it disappears from the warmed portion to the cooler area of the cavity. This indicates that the buoyancy forces are able to overcome the retarding influence of the viscous forces. Figures 11.6a–b display the effect of medium permeability *via* the Darcy number ($Da$) on the local heat transfer rate in terms of the local Nusselt number ($Nu_L$) for different values of the Rayleigh number ($Ra$). As can be observed from these figures, in general, an initial increase

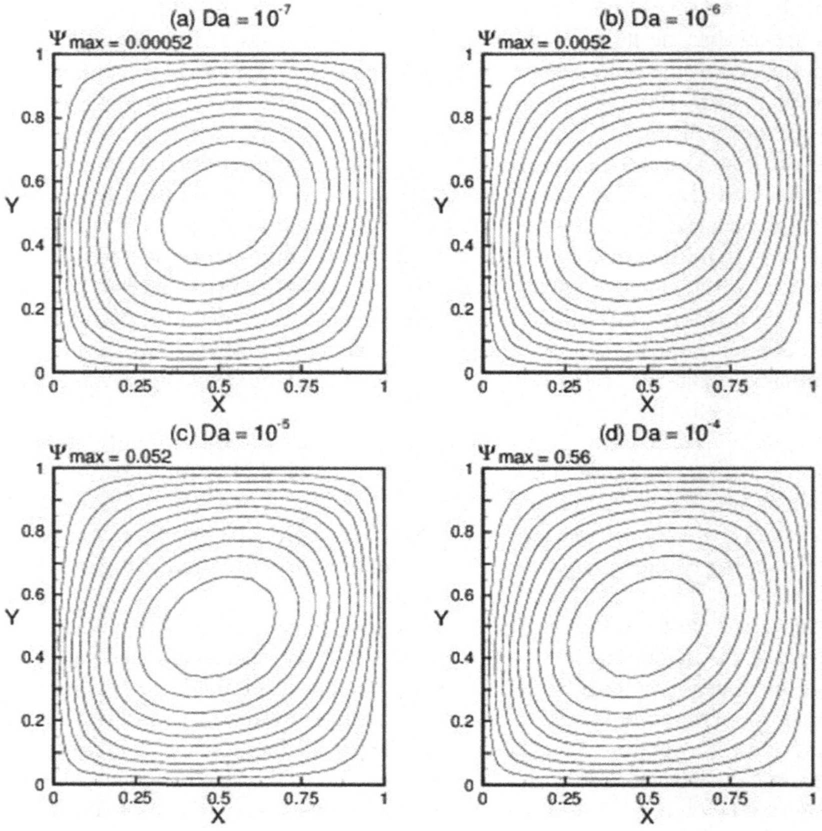


**FIGURE 11.2** Impact of Darcy number (Da) on stream function at Ra = $10^5$.

in *Da* from $10^{-7}$ to $10^{-4}$ increases local heat transfer rates, $Nu_L$, which is based on the fact that increased *Da* is due to an increase in permeability along the direction of the flow. As a consequence, the flow strength increases, and more heat transfer takes place. Secondly, when *Ra* increases from $10^5$ to $10^6$, $Nu_L$ increases and the variation in $Nu_L$ becomes more significant when driving parameter $RaDa \geq 10$. The influence of the Darcy number on the overall heat transfer rate (average Nusselt number, *Nu*) is depicted in Figure 11.7 for different values of *Ra*.

Figure 11.7 shows that the average heat transfer rate, i.e., *Nu*, increases in response to an increase in the medium permeability by varying *Da* from $10^{-7}$ to $10^{-4}$, because increases in permeability allow for greater flow strength in the cavity. As a consequence of this, the above result is expected. An important finding is that the variation in the average heat transfer rate becomes significant when the driving parameter $RaDa \geq 10$.

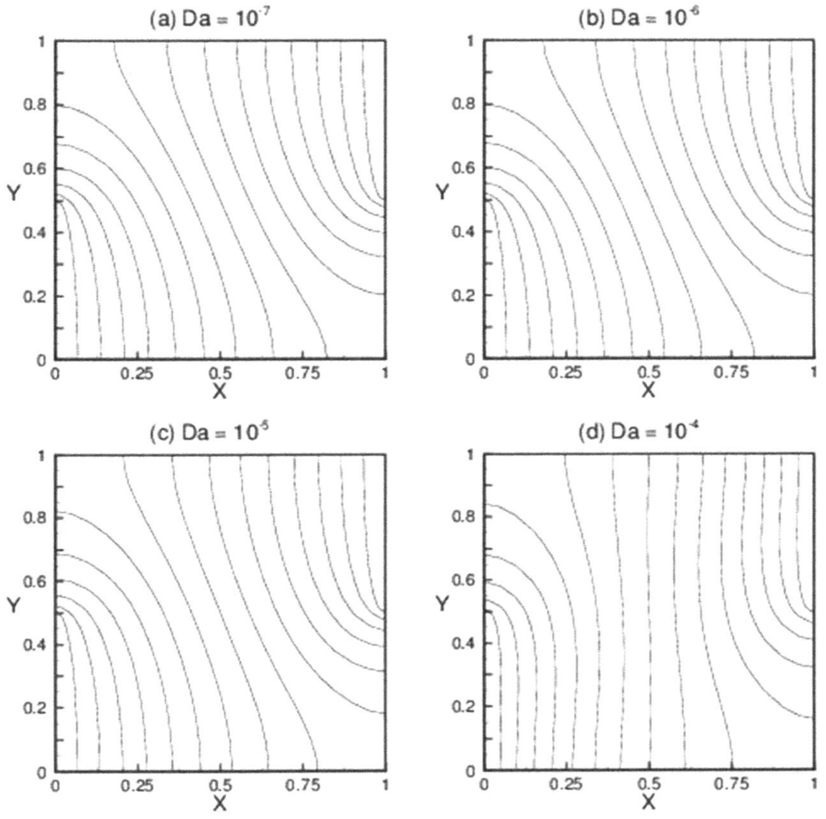


**FIGURE 11.3** Impact of Darcy number (Da) on isotherm contours at $Ra = 10^5$.

The effect of increasing heat source intensity on the stream lines and the isothermal lines is shown in Figure 11.8 and Figure 11.9, respectively, at $Da = 10^{-4}$. Figures 11.8a–b show the effect of the Rayleigh number ($Ra$) on the stream function. It can be seen from these figures that the maximum value of the stream function $\psi_{max}$ increases in response to increasing $Ra$. As the Rayleigh number increases from $10^3$ to $10^6$, the buoyancy-driven circulation inside the cavity also increases, as seen from the greater magnitudes of the stream functions ( Figure 11.8).

Circulation is greater near the wall due to the no-slip boundary conditions. From the same figure (Figure 11.9), it has been calculated that the maximum values of the stream function ($\psi_{max}$) at $Ra$ equal to $10^3$, $10^4$, $10^5$, and $10^6$ are 0.005, 0.05, 0.56, and 4.52, respectively. From this quantitative analysis, it can be concluded that $\psi_{max}$ increases by approximately one order of magnitude in response to an increase of one order of magnitude for the Rayleigh number ($Ra$). The corresponding change in the temperature is shown in Figures 11.9a–b. The influence of the Rayleigh number ($Ra$)

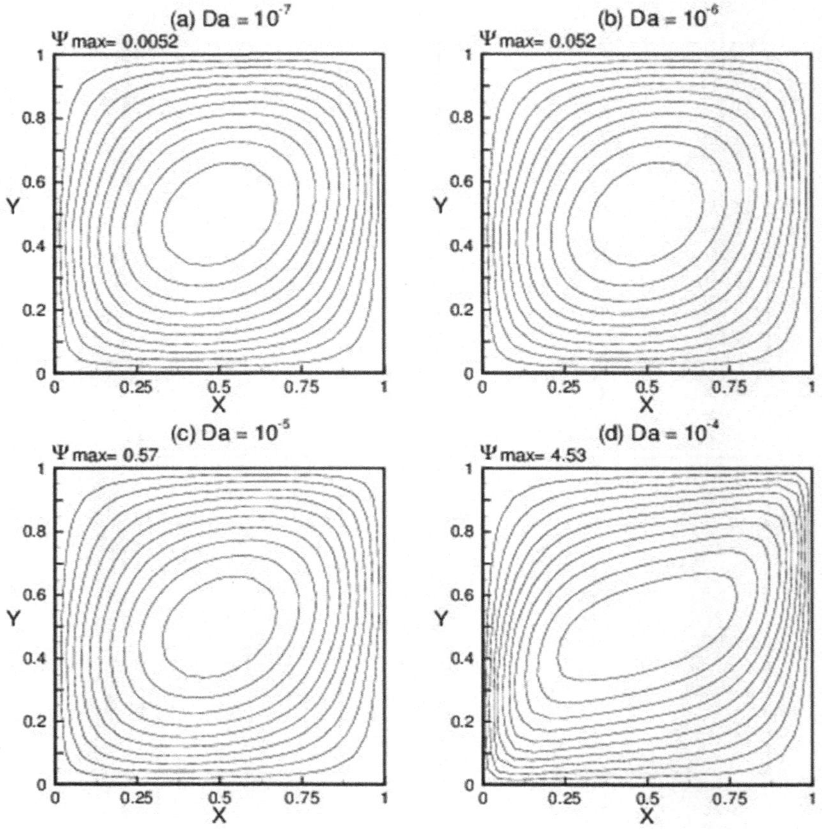


**FIGURE 11.4** Impact of Darcy number (Da) on stream function at $Ra=10^6$.

on the local as well as the average heat transfer rates ($Nu_L$ as well as $Nu$) is plotted in Figures 11.10a–b.

From Figure 11.10a–b, it can be seen that the local heat transfer rate *NuL* increases as the Rayleigh number (*Ra*) increases, whereas the average heat transfer rate (*Nu*) also increases as the Rayleigh number (*Ra*) increases, as can be seen from Figure 11.10b. From the meaning of *Ra*, when different boundaries are fixed, increasing *Ra* suggests an increase in partial heating and cooling, which thus increases the heat convection in the pit. Accordingly, the local heat transfer rate, as well as the average heat transfer rate.

## 11.8 CONCLUSIONS

A thorough mathematical examination of normal convection in a square cavity filled with a permeable medium is introduced. The stream is induced because of

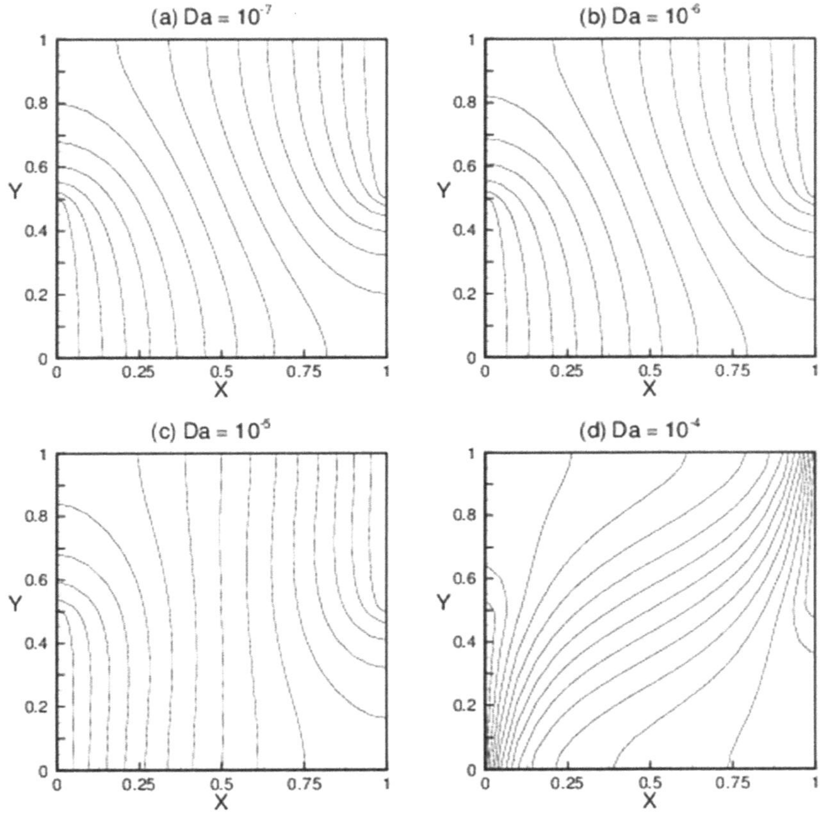


**FIGURE 11.5** Impact of Darcy number (Da) on isotherm contours at Ra = $10^6$.

the steady halfway heating at the left vertical divider and subterranean insect partially cooling at the appropriate vertical divider alongside the adiabatic rest dividers.. In this study, special consideration is given to understanding the impact of medium porosity through the Darcy number and the heat transfer rate, i.e., the Rayleigh number, *Ra*, on the liquid stream set-up with respect to the local and average heat flow rates. The outcomes are described, regarding stream flow $\psi$, temperature $\theta$, and the Nusselt numbers (local Nusselt number, *NuL*, and average Nusselt number, *Nu*).

The following conclusions can be drawn from this study:

- Generally, increasing *Da* from $10^{-7}$ to $10^{-4}$ increases the stream strength in the hole.
- The local and the average heat flow rates as average heat transfer rate (*NuL*, *Nu*) increase in response to increasing medium permeability by changing

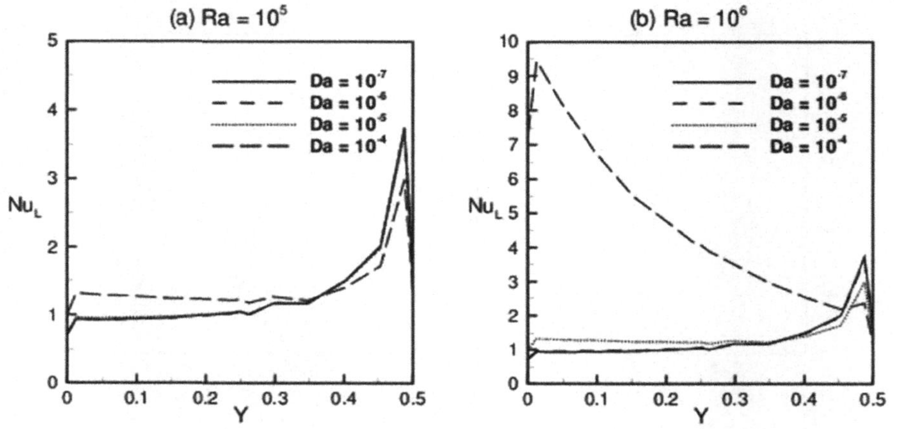


**FIGURE 11.6** Impact of Darcy number (Da) on local heat transfer rate ($Nu_L$) at (a) $Ra = 10^5$ and (b) $Ra = 10^6$.

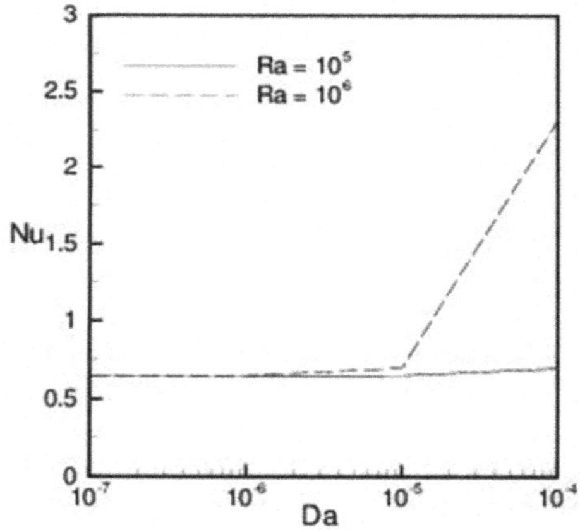


**FIGURE 11.7** Impact of Darcy number (Da) on average heat transfer rate (Nu) for $Ra = 10^5$ and $10^6$.

$Da$ from $10^{-7}$ to $10^{-4}$, because increased permeability allows greater stream flow. This above outcome is typical.

- As the heat transfer rate, for example the Rayleigh number ($Ra$), increases as stream function increases, as a result of which the magnitude of the stream flow increases as the heat flow rates (local as well as average Nusselt numbers) increase.
- The variation in the stream function flow increases as the heat transfer rate when the driving boundary $RaDa \geq 10$.

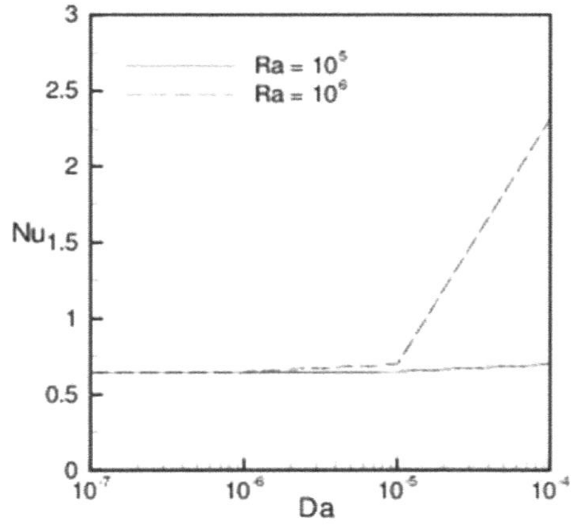


**FIGURE 11.8** Impact of Rayleigh number (Ra) on stream function at $Da = 10^{-4}$.

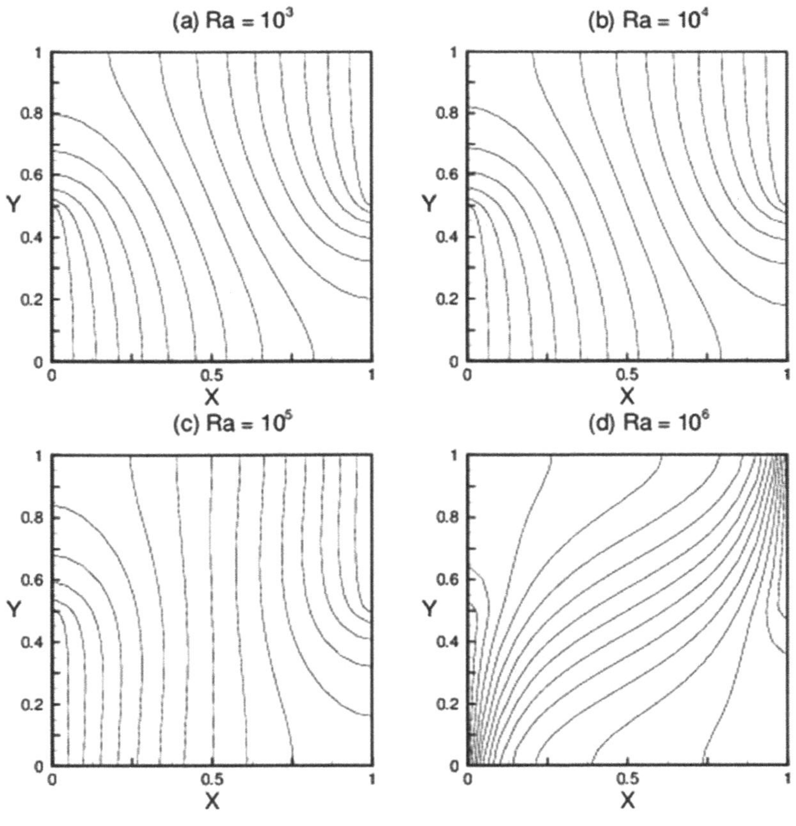


**FIGURE 11.9** Impact of Rayleigh number (Ra) on isotherm contours at $Da = 10^{-4}$.

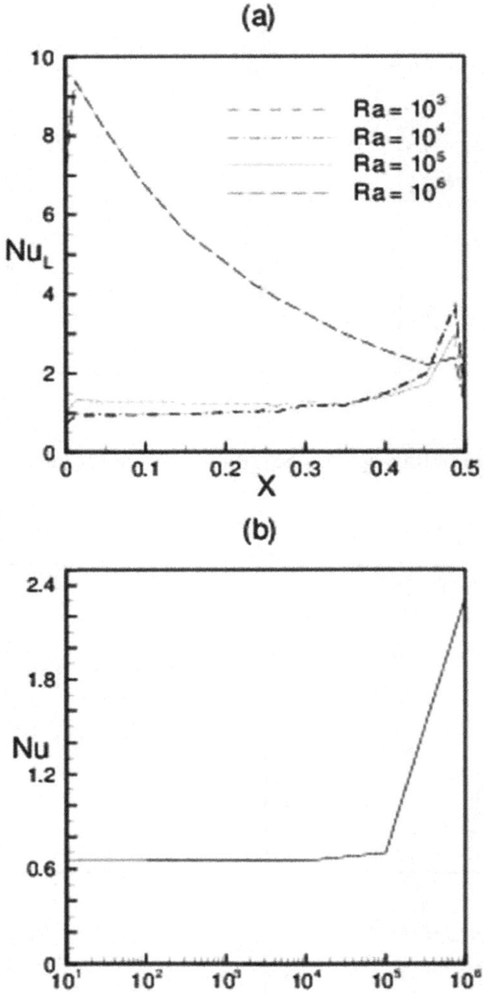


**FIGURE 11.10** Impact of Rayleigh number (*Ra*) on (a) local heat transfer rate ($Nu_L$) and (b) average heat transfer rate (*Nu*) at $Da = 10^4$.

## REFERENCES

Hassinet L. (2019). Numerical study on natural convection in a porous cavity that is partially heated and cooled by sinusoidal temperature at vertical wall. *Journal of Porous Media*, 22, pp. 73–85.

Hsiao S.W., Chen C.K., and Cheng P. (1994). A numerical solution for natural convection in an inclined porous cavity with discrete heat source on one wall. *International Journal Heat Mass Transfer* 37, pp. 2193–2201.

Ingham D.B. and Pop I. (1998). *Transport Phenomenon in Porous Media*. Pergamon, Oxford.

King P.R., Buldyrev S.V., Dokholyan N.V., Havlin S., Lee Y., Paul G., and Stanley H.E. (1999). Applications of statistical physics to the oil industry: Predicting oil recovery using percolation theory. *Physica A* 274, pp. 60–66.

Koponen A., Kandhai D., Hellen E., Alava M., Hoekstra A., Kataja M., Niskanen K., Sloot P., and Timonen J. (1998). Permeability of three dimensional random fiber webs. *Physical Review Letters* 80, pp. 716–719.

Lai F.C. and Kulacki F.A. (1991). Experimental study of free and mixed convection in horizontal porous layers locally heated from below. *International Journal of Heat Mass Transfer* 34, pp. 525–541.

Lai F.C., Choi C.Y., and Kulacki F.A. (1990). Free and mixed convection in horizontal porous layers with multiple heat sources. *Journal of Thermophysics and Heat Transfer* 4, pp. 221–227.

Lakhal E.K., Hasnaoui M., Vasseur P., and Bilgen E. (1995). Natural convection in a square enclosure heated periodically from part of the bottom wall. *Numerical Heat Transfer: Part A Applications* 27, pp. 319–333.

Lauriat A. and Prasad V. (1989). Non-darcian effects on natural convection in a vertical porous enclosure. *International Journal of Heat and Mass Transfer* 32, pp. 2135–2148.

Nader B.C., Brahim B.B., and Taieb L. (2007). Influence of thermal boundary conditions on natural convection in a square enclosure partially heated from below. *International Communications in Heat and Mass Transfer 34*, pp. 369–379.

Natesan S. (2017). Heat enhancement of uniformly/linearly heated side wall in a square enclosure utilizing alumina-water nanofluid. *Computational Thermal Sciences: An International Journal*, 9, pp. 227–241.

Nield D.A. and Bejan A. (2006). *Convection in Porous Media*. Springer, New York.

Robillard L. (1988). Multiple steady states in conned porous medium with localized heating from below. *Numerical Heat Transfer: Part A-Applications* 13, pp. 91–110.

Saeid N.H. and Pop I. (2005). Natural convection from a discrete heater in a square cavity filled with a porous medium. *Journal of Porous Media* 8, pp. 55–63.

Sarris I.E., Lekakis I., and Vlachos N.S. (2004). Natural convection in rectangular tanks heated locally from bellow. *International Journal of Heat and Mass Transfer* 47, pp. 3549–3563.

Sharif M.A.R. and Mohammad T.R. (2005). Natural convection in cavities with constant flux heating at the bottom wall and isothermal cooling from the sidewalls. *International Journal of Thermal Sciences* 44, pp. 865–878.

Turgoklu H. and Yucel N. (1995). Effect of heater and cooler locations on natural convection in square cavities. *Numerical Heat Transfer: Part A Applications* 27, pp. 351–358.

Zhao F.Y., Liu D., and Tang G.F. (2008). Natural convection in a porous enclosure with a partial heating and salting element. *International Journal of Thermal Sciences* 47, pp. 569–583.

# Index